PHILOSOPHY AND COGNITIVE SCIENCE:
CATEGORIES, CONSCIOUSNESS, AND REASONING

PHILOSOPHICAL STUDIES SERIES

VOLUME 69

PHILOSOPHY AND COGNITIVE SCIENCE: CATEGORIES, CONSCIOUSNESS, AND REASONING

Proceedings of the Second International Colloquium on Cognitive Science

Edited by

ANDY CLARK

Philosophy/Neuroscience/Psychology Program,
Department of Philosophy,
Washington University,
St Louis, Missouri, U.S.A.

JESÚS EZQUERRO

Department of Logic and Philosophy of Science,
Institute for Logic, Cognition, Language and Information (ILCLI),
Universidad del Pais Vasco/EHU,
San Sebastian, Spain

and

JESÚS M. LARRAZABAL

Department of Logic and Philosophy of Science,
Institute for Logic, Cognition, Language and Information (ILCLI),
Universidad del Pais Vasco/EHU,
San Sebastian, Spain

Kluwer Academic Publishers

Dordrecht / Boston / London

A C.I.P. Catalogue record for this book is available from the Library of Congress.

ISBN 978-90-481-4710-6

Published by Kluwer Academic Publishers,
P.O. Box 17, 3300 AA Dordrecht, The Netherlands.

Kluwer Academic Publishers incorporates
the publishing programmes of
D. Reidel, Martinus Nijhoff, Dr W. Junk and MTP Press.

Sold and distributed in the U.S.A. and Canada
by Kluwer Academic Publishers,
101 Philip Drive, Norwell, MA 02061, U.S.A.

In all other countries, sold and distributed
by Kluwer Academic Publishers Group,
P.O. Box 322, 3300 AH Dordrecht, The Netherlands.

Printed on acid-free paper

CONTENTS

viii

INTRODUCTION

PHILOSOPHY AND COGNITIVE SCIENCE: CATEGORIES, CONSCIOUSNESS, AND REASONING

> *The individual man, since his separate existence is manifested only by ignorance and error, so far as he is anything apart from his fellows, and from what he and they are to be, is only a negation.*

> Peirce, Some Consequences of Four Incapacities. 1868.

For the second time the International Colloquium on Cognitive Science gathered at San Sebastian from May, 7-11, 1991 to discuss the following main topics:

> Knowledge of Categories
> Consciousness
> Reasoning and Interpretation
> Evolution, Biology, and Mind

It is not an easy task to introduce in a few words the content of this volume. We have collected eleven invited papers presented at the Colloquium, which means the substantial part of it. Unfortunately, it has not been possible to include all the invited lectures of the meeting.

Before sketching and showing the relevance of each paper, let us explain the reasons for having adopted the decision to organize each two years an international colloquium on Cognitive Science at Donostia (San Sebastián). First of all, Cognitive Science is a very active research area in the world, linking multidisciplinary efforts coming mostly from psychology, artificial intelligence, theoretical linguistics and neurobiology, and using more and more formal tools. We think that this new discipline lacks solid foundations, and in this sense philosophy, particularly knowledge theory, and logic must be called for. This is why a Department like ours is engaged in the organization of these Colloquia. Secondly, we also think that it is very important to improve all the research fields involved in this area taking especially into account a European framework, where there are many groups producing innovative results but without a sufficient connection among them, which means that the intellectual benefits do not amount to the level required. It seems very clear to everybody that the emergence of an integrated science of cognition needs a strong collaboration of the research teams, given that the task is twofold: on the one hand, Cognitive Science must rethink in its own perspective the old philosophical and psychological problems connected to the processes of cognition and action, and on the other hand, this Science has to answer the new questions arisen inside in order to explain rational behavior in the most general case.

The improvement of a general program in Cognitive Science, centered in the analysis of the relation between Language and Cognition, has been one of the main goals of the Department in the last years. Thus, an important number of Ph. D. Theses have been carried out in the frame of a Doctorate Program on "Foundations of Logic, Computer Science and Cognitive Science", and currently new doctoral theses are in realization in a succeeding program called "Logic, Language, Science".

The creation of the Institute for Logic, Cognition, Language and Information (ILCLI) at the University of the Basque Country has been crucial for strengthening research in Cognitive Science. Indeed, this is one of the four main research areas of the Institute. The other three (Logic; Foundations of Computer Science and Artificial Intelligence; and Language, Information, and Communication) are clearly related to it. Before turning to introduce the contents of this volume, let us mention that a Diploma in Cognitive Science and a Master of Language, Information, and Communication are offered by the Institute.

No doubt, we live an exciting moment for developing a science of cognition. We know more and more about the psychological complexity of mental states and so we are more and more aware of the subtle processes governing rational action. In the same way Artificial Intelligence, and particularly Distributed Artificial Intelligence, is offering a better understanding of what rational behavior means, through more and more realistic models. It seems that the use of different representation paradigms and instrumental techniques can at least converge in the solution of some local problems, which have resisted previous approaches.

However, it is evident that many theoretical issues in Cognitive Science need a conceptual clarification in order to avoid important difficulties on the road. Conceptual clarification is not an easy task as far as it is not easy at all to identify the relevant philosophical points involved in the very foundations of Cognitive Science. Triviality is the main risk of philosophers intending to ground the foundations of any science. Moreover, the analysis of the philosophical problems demands a deep knowledge of all the features of cognition and action. That means that philosophy can be one of the guides of Cognitive Science, if it is basically understood as epistemology, far away from a "master-slave" relation as unfortunately has been the case in traditional philosophy.

The title "Philosophy and Cognitive Science: Categories, Consciousness and Reasoning" tries to reflect the main issues addressed in the volume and gives some idea about its structure. Chapters 1 to 8 deal with mental contents from several perspectives. Some keywords can help to link the different contributions: narrow and wide content; internalism and externalism; relations between mind, language and reality; mental content and consciousness; and consequences on these topics derived from the connectionist style of representation. Finally, Chapters 9 to 11 deal with reasoning under different perspectives.

The first paper in this volume is by Martin Davies on "Externalism and Experience". There are two classical papers in the literature for arguments purporting to establish externalism: one is Hilary Putnam's *The Meaning of 'Meaning'*, and the other is Tyler Burge's *Individualism and the Mental*. In the former essay, Putnam proposed the distinction between *narrow* and *wide* content. It is well-known that during the last years there has been a hot controversy related to the above notions: Internalism *versus* Externalism (often referred to as "Individualism *versus* Anti-individualism"). Internalism is tied to narrow content, thus individuating content inside the notional world of the subject. On the contrary, externalism claims that, when individuating content, environmental and sociolinguistic factors must be taken into account as well. Davies enters in this controversy, although in a oblique way. On the one hand, he defends externalism for the contents of perceptual experience, but, as he notes, "perceptual content is a distinct kind of content, different from belief content". This fact involves that "arguments for externalism about belief content cannot automatically be transposed into arguments for externalism about perceptual content". His claim is in favour of externalism. Nevertheless, there is a problem with this claim: the individualist has the advantage that her position respects local supervenience, and this is a real challenge that externalists must face up. Davies deals with this problem by noting that supervenience claims may vary in strength along *modal* dimensions. Such dimensions can range from 'across worlds' (WW') and 'across subjects' (XY) to 'within a world' (WW) and 'within a subject' (XX). The modally strong supervenience (XYWW') is at issue in the debate between individualism and externalism. Davies takes instead a modally modest supervenience in his defence of externalism for perceptual content.

The next two papers (William Ramsey and Andy Clark) are concerned with connectionist accounts of concepts. The way categories are acquired and organized in connectionist systems has important consequences for their cognitive abilities and for the kind of theories they can support. Neural networks learn concepts by a long training with numerous different instances of them. The result is not a tidy, well defined class but a superposition of patterns which bear a resemblance. Every new instance of the concept is not contrasted with a definition to be judged as a member of the category. Instead, the distance from the encoded patterns is measured, and that amounts to an appraisal of its resemblance to the rest of instances. From this style of categorizing and representing, our two authors draw different but complementary consequences: one concerned with the impact of Cognitive Science on Philosophy and the other with the impact of Philosophy on Cognitive Science.

William Ramsey, in his contribution on "Conceptual Analysis and the Connectionist Account of Concepts" explores the consequences that this approach can carry out on the way we represent concepts, and so, on what is derived for conceptual analysis as a philosophical task. The author considers that there is a real clash between the most popular connectionist theories and the basic assumptions driving conceptual analysis, say, "the search for precise definitions specifying necessary and sufficient conditions for abstract notions". Ramsey makes a conditional statement: if connectionist accounts prove correct, then we will have to revise such an assumption. Ramsey realizes that distributed connectionist models of any variety judge category membership by similarity to a prototype. Notions of necessary and sufficient conditions do not fit here, but this account has psychological support and we can invoke the work of Rosch *et al.* to provide a cognitive account of this phenomenon. Thus, the prospects for conceptual analysis under connectionist representational theory are at odds with classical views. He notes that this clash can only be avoided by relaxing one of the criteria grounding conceptual analysis, namely, in his own words, "abandoning simplicity in order to accommodate all intuitions may be a misguided strategy, and a more promising approach might be the one of preserving simplicity and recognizing that we can't capture all intuitive judgements".

On the contrary, the flexibility for categorization judgements that results from distributed connectionist architectures raises some problems under Andy Clark's analysis on "Of Norms and Neurons". To begin with, Clark argues in this paper against himself, in the sense that he changes his mind in relation to a radical position stating that philosophy has nothing to say about the internal structure of intentional systems. In the present paper, Clark thinks that the view that any system exhibiting behavioural patterns appropriately complex can count as a genuine believer needs to be revisited, because of its inadequacy to account for some normative requirements. The author posits two philosophical constraints on intentional systems (the Generality Constraint, and the Requirement of Normative Depth) assuming that they apply to the innards of a cognitive agent and constitute a minimal organizational structure that rules out behavioural simulations such a Giant

Look-up Tables. Next, he considers them in the case of a connectionist network. The problem, as anticipated, lies in the way connectionist systems acquire their categories: statistical learning is not suitable to grasp the constitutive features of concepts. Clark concludes that only some advanced forms of connectionist models can meet the above two constraints, and furthermore that it is also the case for some non-representational approaches *à la* Brooks (remember, "Intelligence without Representation").

Keith Lehrer's paper is on "Skepticism, Lucid Content and the Metamental Loop". Lehrer is well-known in Theory of Knowledge, and in this contribution he is particularly interested in a topic that has received little attention in contemporary Cognitive Science, say, the special kind of knowledge that we have of the contents of our own thoughts, what he calls *lucid contents*. The reason of Lehrer's interest in this issue is that so far Cognitive Science has focused mainly on representation, functional role and so forth, but for Lehrer knowledge is more than representation, given that agents can have a representation and not know that such a representation is correct. It must be remarked that for him knowledge is basically undefeated personal justification, and that this is the account he applies to explain lucid content and the implications of such knowledge. Following Lehrer, we say that someone knows *p* just in the case she is able to meet all skeptical objections that count against *p*, and possible errors in her background acceptance system would not defeat the personal justification she has for maintaining that *p* is the case. However, internal justification is not enough for knowledge, truth is also needed. Lehrer introduces truth by means of the "non-defeatability clause": corrections of errors in the acceptance system must not defeat the personal justification. This is when the metamental plays a role: knowledge requires the metamental in stemming from the second order process of evaluating representations and from the correctness of that evaluation. Nevertheless, this account has the threat of infinite regress, and the only way to escape is to posit what he calls the *metamental loop*. Its explanation is briefly provided in the following words: "My primary thesis is that metamental processing and evaluation convert mental representation into knowledge and mental activity into lucid content".

The next contribution is by John Perry on "Evading the Slingshot". Really that had been evaded long time ago by Perry (remember some joint work by Barwise and Perry before *Situations and Attitudes*). The idea for dismantling the "slingshot" was based on the rejection of the principles called substitution and redistribution, in such a way that it could be defended the view that sentences do not designate truth-values, but stand for complexes of objects and relations, as Situation Theory claims. What is new in this paper is a thoroughgoing discussion of Ken Olson's arguments in his book *An Essay on Facts* about the view considering Barwise and Perry's approach as a structuralist one. According to the structuralist conception, facts are individuated by sequences of properties and objects. Instead, according to what Olson calls the "existential" conception, facts are identical if they necessarily co-exist. While the latter supports redistribution, the former allows us to claim that redistribution movements aren't content preserving. This inclines Perry to adhere to the structuralist account, given that it allows one to block the slingshot. The problem is that, in this case, we must face Olson's objection that the structuralist conception improperly mixes metaphysics and syntax, and that, in turn, it would be susceptible to a sort of metaphysical slingshot. Specifically, the problem with the structuralist account of facts is the one of the "block universe": you cannot individuate a fact without taking all the universe with you. The solution proposed by Perry, however a tentative one, is to resort to *situations* as the basics of the world. This way, situations may be used to determine facts.

The sixth paper is by Alfonso García Suárez on "Reference without Sense: An Examination of Putnam's Semantic Theory". It deals directly with the conclusions Putnam drew in his famous *The Meaning of 'Meaning'*. García Suárez tries to show two things: first, that even admitting the soundness of the hypothesis of the division of linguistic labour, it does not state a serious challenge to a traditional theory of meaning, and second, that the thesis of indexicality is not correct. It is important to distinguish two possible scenarios in Twin-Earth cases, in order to understand what García Suárez is trying to convey. The first one takes place in 1750, before water was discovered to be H_2O. The second after such a discovery was made. Let us assume that, in the latter scenario, an Earthian, having no idea of Daltonian chemistry, and ignoring thus that water is H_2O, has the chance to travel to Twin-Earth and gets in contact with twin-water. Of course, our Earthian also ignores that twin-water is XYZ. Nevertheless, she is aware that Earth experts know how to distinguish water from twin-water. In this case, there is a difference in the person's conceptual knowledge, given that she knows, by asking experts, how to pick up successfully water instead of water and twin-water. For this reason, the author thinks that Putnam's thought experiment counts just against Methodological Solipsism, but not against traditional theory, that could easily accommodate communal knowledge and the division of linguistic labour. On the contrary, the former scenario presents harder problems to a Fregean theory of meaning. Given the lack of resources for picking up water as H_2O, a Fregean would be forced to admit that someone thinking, for example, "water is wet" must include as referents both water and twin-water. For this reason Putnam concluded that meanings understood in the narrow sense, say, as what mind grasps in comprehension, cannot determine extension, and thus meaning has to be a wider notion. But in this case, if we allow a notion of meaning broad enough as to include extension (remember, Putnam's cluster theory), then the very idea that meaning determines extension becomes trivial and unexplained. The point at issue is, García Suárez contends, that the notion of intension (or sense) is an epistemic notion. Its role is played in Putnam's cluster theory by stereotypes, but stereotypes cannot determine extension. Therefore, the prospects of making sense of the assumption that meaning determines extension turn out to be very difficult, because we must confront the following dilemma: either to embrace Putnam's cluster theory, in which case we were doing a Pickwickian move, or to come back to a Fregean theory, that conflates an epistemic notion of "sense" with a semantic one (say, "sense" as a route towards reference). García Suárez's positive claim is to suggest that there is still a possible way to escape from the above dilemma, even for a supporter of Putnam's views, namely, by trying to reconcile the above two functions of "sense". For if we were to allow a specification like the following one: *"Water is whatever is identical in nature with THIS* (indexically identified) *stuff, whatever its nature may be"*, a defender of the cluster's theory would have no objection to the idea that such a specification partially fixes the extension of the term "water". The problem is, our author claims, that such a move would amount to convert Putnam's theory into a modified version of Fregean semantics.

   In addition to Putnam's cases, Frege's cases constitute another hard problem that any theory of language and mind must resolve. Juan J. Acero's contribution on "Attitudes, Content and Identity: A Dynamic View" deals with this issue. At first glance, Frege's puzzle states a problem for causal-informational theories, because within their frame it is not easy to differentiate between the propositions expressed by:

   (1)   Anakin Skywalker does not serve the emperor

   (2)   Darth Vader does not serve the emperor,

in spite of their obvious informative difference. Acero notes that although attempts to answer the above question realize that Frege's puzzle poses a problem concerning the nature of beliefs, they have failed so far because it has usually been assumed that mental states have a fixed content. Another intimately related problem is the so called Kripke's puzzle. Kripke proposed to consider the case of a French speaker, Pierre, that when living in Paris becomes to believe *"Londres est jolie"* (London is pretty), but that some time later, while living in an unattractive part of London realizes that "London is not pretty". Of course, Pierre has learnt English by the 'direct method' and thus he does not know that both names refer to the same city. In this case, it looks as though Pierre endorses two contradictory beliefs. Acero thinks that the problems brought about by the above puzzles cannot be solved unless we give up the assumption that contents are static. In a nutshell, our author claims that contents are dynamic, and depend not only on the situation or state of affairs subjects are linked to, but on what else they believe at each moment as well. The relevant aspect of the paper is his claim defending that the dynamic viewpoint is compatible with what he calls (in singular) the New Theory of Reference in Philosophy of Language and Externalism in Philosophy of Mind. It is worth mentioning his way of dealing with belief dynamics by means of notional worlds.

We have indicated how W. Ramsey, in Chapter 2, vindicates the notion of "prototype" in conceptual analysis, as a result of the representational style of connectionist systems. Later, in chapters 4 to 7, the authors contributing to the present volume have dealt with several aspects relating semantics and mental representation. Now, Peter Gärdenfors' paper on "Conceptual Spaces as a Basis for Cognitive Semantics" can be considered as a bridge between these perspectives. To begin with, Gärdenfors draws a sharp distinction between realistic semantics and cognitive ones. Usually realistic semantics define meaning in terms of truth-conditions. This is particularly the case of intensional semantics, considering possible worlds. On the contrary, cognitive semantics define meaning by means of cognitive structures, in such a way that the relation between meaning and the external world(s) becomes secondary. Gärdenfors' proposal consists in establishing an ontological framework for cognitive semantics. This framework is provided by conceptual spaces. Our author thinks of his theory of conceptual spaces as embodying several claims. In addition to the thesis that meaning comes before truth (meaning is considered as "conceptualization in a cognitive model"), Gärdenfors endorses the following commitments: "cognitive models are mainly perceptually determined"; "semantical elements are based on spatial or topological objects"; "cognitive models are primarily image-schematic, and image-schemas are transformed by metaphoric and metonymic operations"; "semantics is primary to syntax and partially determines it"; and "concepts show prototype effects". Our author ends by showing how this kind of semantics can be useful to understand metaphors and prototype effects of concepts.

The last three chapters are specifically devoted to the study of different modalities of reasoning. The first one explores the differences in reasoning according to the representational medium in which it is performed. The second is concerned with a particular object of reasoning, relations of cause-effect. The third tries to outline principles of reasoning when it is performed under conditions of uncertainty.

Chapter 9 is by Keith Stenning on "The Cognitive Impact of Diagrams". This proposal must be included in a more general theory, namely, a theory of "graphical communication", which aims to extend the procedures used in linguistic communication. In fact, such a theory intends to ground many different practices of message transmission. The assumption behind Stenning's proposal is that expressive power varies as far as the media to convey the message differ. In this sense, graphical representations are quite specific and, contrary to what happens in natural language, cannot avoid expressing some types of information. The important point is the kind of abstractions and conventions that guide the construction and interpretation of graphics. The author considers particularly Euler's Circles, which were designed for teaching syllogistics, given the fact that this system has been used for several psychological models of commonsense reasoning. Stenning argues that the representational style of Euler's Circles has been often misunderstood both by those who make use of them and by those who reject them on the basis of their intractability (remember Johnson-Laird), because they have not taken into account the opportune interpretation conventions. Careful study of graphical representation helps to cast light on the related areas of analogical reasoning and the implementation of fully expressive languages into less expressive ones. Limitations of logical expressiveness seem to be needed to account for the way human reasoning is performed. This is the task Stenning tries to carry out with Euler's Circles, a case study that can be quite easily generalized to other graphical systems of representation.

Kurt Konolige focuses on commonsense causation in a paper entitled "What's Happening? Elements of Commonsense Causation". He is concerned with establishing a theory of causal reasoning in the frame of what is called commonsense reasoning in Artificial Intelligence, assuming "normal conditions" for the validity of that reasoning. These normal conditions are assumed by "default", and so his proposal aims at an integration of causal and default reasoning, using default causal nets to represent the causal connections among propositions in the domain of representation. In his own words, "the theory ... is a contribution in the application of default reasoning techniques to a commonsense reasoning domain. It develops the concept of an irreducible causation relation, and joins to this the ability to make assumptions about normal conditions. The interaction between causation and assumption is complex; the most interesting fact is in the inference processes of prediction and explanation...".

The last paper in this volume is by Jeff Paris and Alena Vencovska on "Principles of Uncertain Reasoning". This contribution is devoted to a formulation of some principles for uncertain reasoning in general, and particularly for a probability logic in commonsense reasoning. The principles they propose from an idea of rationality are the following ones: Equivalence, Irrelevant Information, Continuity, Open Mindedness, Indifference, Obstinacy, Relativisation, Independence, and Atomicity. An idea of

Calibration is added in order to fulfil a requirement of accuracy with a rather strong notion of correctness.

Some words for conclusion. It is clear that the papers collected in this volume represent different points of view and come from different research fields in Cognitive Science. Philosophy, logic, artificial intelligence, psychology, semantics, and mathematics are present. Obviously that is not all, but it is a substantial part of what is done in Cognitive Science. Our purpose was to offer both a perspective of the main topics of the conference and at the same time the most recent multidisciplinary results inviting further research on Categories, Consciousness, and Reasoning.

Acknowledgements. The organization of this Second Colloquium was sponsored by the Secretary of State of Education and Science (Ministerio de Educación y Ciencia) of the Spanish Government, the Secretary of Culture of the Basque Government, the Secretary of Culture of the County Council of Gipuzkoa (Gipuzkoako Foru Aldundia), the Council of Kutxa (Saving-Bank of Gipuzkoa), and the President of the University of the Basque Country. Many thanks to them, on behalf of the Department of Logic and Philosophy of Science of that University, which was the organizer of the conference. Fernando Martínez and Agustín Vicente helped us by correcting proofs and preparing the Subject's and Author's Index, and Xabier Arrazola and Kepa Korta solved our technical troubles. We would like to thank Kluwer Academic Publishers for their patience in helping to put this volume together.

The editors

MARTIN DAVIES

EXTERNALISM AND EXPERIENCE

I. INTRODUCTION

In this paper, I shall defend externalism for the contents of perceptual experience. A perceptual experience has representational properties; it presents the world as being a certain way. A visual experience, for example, might present the world to a subject as containing a surface with a certain shape, lying at a certain distance, in a certain direction; perhaps a square with sides about 30 cm, lying about one metre in front of the subject, in a direction about 20 degrees to the left of straight ahead.

There are two views that we might take about these representational, or semantic, properties of experiences. On the one hand, we might hold that the content of a perceptual experience is just the content of the judgement that the subject would make if he or she took the experience at face value. In that case, perceptual content is the same kind of content as the content of judgement and belief; and externalism about perceptual content is just the same as externalism about belief content. On the other hand, we might hold that perceptual content is a distinct kind of content, different from belief content. In that case, arguments for externalism about belief content cannot automatically be transposed into arguments for externalism about perceptual content.

I shall be adopting this second view: externalism concerning perceptual content requires separate argument. That argument comes in the sixth section of this paper. The second section offers a little more clarification of what is distinctive about perceptual content, while the third section characterises the externalist's claim. The fourth section sets out in more detail what is required of an externalist argument, and the fifth section explains a dilemma that the externalist is liable to face - a dilemma that I try to avoid in the sixth section. The seventh and final section raises the question whether externalist perceptual content should be conceived of as a representational superstructure erected upon a sensational substrate.

1

A. Clark et al. (eds.), Philosophy and Cognitive Science, 1–33.
© 1996 *Kluwer Academic Publishers.*

II. PERCEPTUAL CONTENT

On the view that I am adopting here, the perceptual content of an experience is a kind of *non-conceptual* content. What this means is that a subject can have an experience with a certain perceptual content without possessing the concepts that would be used in specifying the content of that experience. Indeed, the philosophical category of perceptual content applies equally to the experiences of normal adult human beings, who are deployers of a rich repertoire of concepts, and to the experiences of human infants and certain other creatures, who arguably are not deployers of concepts at all. Enjoying experiences with perceptual content does not require the possession of concepts; *a fortiori*, it does not require the employment of such concepts as may be possessed by the experiencer.

Because the perceptual content of an experience is a kind of non-conceptual content, it must be distinguished from the content of any judgement that might be made if the experience is taken at face value. An experience may present the world to a subject as containing something square in front of her; and the subject may take that experience at face value and judge that there is, indeed, something square in front of her. Making the judgement requires possession and employment of the concept of being square; but merely undergoing the experience does not.

This is not to deny that there is a close connection between the non-conceptual content of experiences, and possession of observational concepts, such as the concept of being square. Thus, Christopher Peacocke says (1989, p. 5):

> We can consider the case of a possession condition for a relatively observational concept. It is plausible that such a possession condition will link mastery of the concept in question to the nonconceptual representational contents of the thinker's perceptual experience.

Possession of those concepts requires a certain answerability of judgements to the perceptual content of experiences. As Colin McGinn says (1989, p. 60), 'to have the concept *square* just is to apply it on the basis of experiences as of square things'. But it is the notion of perceptual content (of experiences *as of* . . .) that comes first in the order of philosophical explanation, and is then appealed to in an account of what it is to possess a concept such as the concept of being square.

To help fix the idea of non-conceptual perceptual content, we can note here that Peacocke offers *scenario content* as just such a kind of content. Here is what he says to introduce scenario content (1992, pp. 61-2):

> I suggest that one basic form of representational content should be individuated by specifying which ways of filling out the space around the perceiver are consistent with the representational content's being correct. The idea is that the content involves a spatial *type*, the type being that under which fall precisely those ways of filling the space around the subject that are consistent with the correctness of the content.

And here is the point that scenario content is non-conceptual (1992, p. 63):

> There is no requirement at this point that the conceptual apparatus used in specifying a way of filling out the space be an apparatus of concepts used by the perceiver himself. Any apparatus we want to use, however sophisticated, may be employed in fixing the spatial type, however primitive the conceptual resources of the perceiver with whom we are concerned.

We can also use the notion of scenario content to illustrate two further features of perceptual content. The first is that perceptual content abstracts from the identities of the particular individual objects that may be perceived: perceptual content is *not object-involving*. The second is that, despite the fact that it is not object-involving, perceptual content determines correctness conditions: perceptual content is *fully representational*. (We should note that, in Peacocke's account, a further layer of perceptual content - *protopropositional content* - is introduced, to mark distinctions between perceived axes of symmetry, for example (1992, p. 77). Possession of the concept of being square then requires answerability of judgements - that a presented object is square - to the protopropositional content, rather than simply the scenario content, of experiences. This is an important complication. But protopropositional content retains the three features that we have noted: it is non-conceptual, it is not object-involving, and it is fully representational.)

Finally in this section, we can achieve some further clarification of the present use of the idea of non-conceptual perceptual content by looking back at the introduction of the notion by Gareth Evans (1982). In fact, Evans introduces the notion of non-conceptual content in two slightly different contexts: unconscious information processing, and perceptual experiences. In the first case, he says (1982, p. 104, n. 22):

> When we attribute to the brain computations whereby it localizes the sounds we hear, we *ipso facto* attribute to it representations of the speed of sound and of the distance between the ears, without any commitment to the idea that it should be able to represent the speed of light or the distance between anything else.

The point here is that when we talk about 'the information-processing that takes place in our brains' we attribute representational properties - contents involving speed and distance, for example - to states of the brain, without any requirement that the conditions for the possession of concepts - for possessing the concept of speed or the concept of distance, for example - should be met. In the second case, he says (1982, pp. 226-7):

> In general, we may regard a perceptual experience as an informational state of the subject: it has a certain *content* - the world is represented a certain way - and hence it permits of a non-derivative classification as *true* or *false*. . . .
> The informational states which a subject acquires through perception are *non-conceptual*, or *non-conceptualized*.

And, in a remark that appears to apply to both cases, he recommends that we 'take the notion of *being in an informational state with such-and-such a content* as a primitive notion for philosophy, rather than ... attempt to characterize it in terms of belief' (1982, p. 123).

Now, the notion of non-conceptual content does, indeed, have an important philosophical role to play in both these contexts. In the first case - that of unconscious information processing - it can figure in an account of *tacit knowledge* of rules, for example (Davies, 1989; Crane, 1992b, pp. 156-7). But it is, of course, the second case that primarily concerns us here; and the question that we need to ask is this. What, according to Evans, distinguishes the second case - the non-conceptual content of perceptual experiences - from the first case - the non-conceptual content of unconscious informational states?

We can see the answer to this question very clearly, if we consider what Evans says about the spatial element in the non-conceptual content of perceptual states. First, spatial content requires links to spatial behaviour: 'we must say that having spatially significant perceptual information consists at least partly in being disposed to do various things' (1982, p. 155). But the connections between informational states and appropriate behaviour could be in place even while there was no conscious subject, and so no perceptual *experiences*. So, for the non-conceptual content of perceptual experiences, something more is required (1982, p. 158):

> we arrive at conscious perceptual experience when sensory input is not only connected to behavioural dispositions . . . but also serves as the input to a *thinking, concept-applying, and reasoning system.*

Crucially, Evans equates a conscious subject with a thinking subject; he equates the consciousness of perceptual experiences with a kind of accessibility of the non-conceptual content of those experiences to the system of conceptualised judgement and belief formation (1982, p. 227):

> In the case of [concept-exercising and reasoning] organisms, the internal states which have a content by virtue of their phylogenetically more ancient connections with the motor system also serve as input to the concept-exercising and reasoning system. Judgements are then *based upon* (reliably caused by) these internal states; when this is the case we can speak of the information being 'accessible' to the subject, and, indeed, of the existence of conscious experience.

For Evans, then, we only find experiences with perceptual content where we also have a thinker. Where there is no thinker, there is no conscious experience; and the perceptual states with non-conceptual content are like the informational states that enable a blindsight patient to 'guess' correctly the direction of a light source (1982, p. 158).

This element of Evans's account is not preserved in the use that I am making of the idea of non-conceptual perceptual content. I would not, myself, impose such a strict standard for perceptual experience, since it closes some questions about the notion of consciousness which I would prefer to leave open. In particular, it seems to be implicit in Evans's account that the *phenomenal* consciousness of perceptual experiences is best understood as a kind of *access* consciousness. (For this distinction, see Block 1990, 1991, 1992, 1993, 1995; and for further discussion, see Davies and Humphreys, 1993.) As our understanding of consciousness improves, we may come to regard that as the correct position. But, pending such an improved understanding, I would prefer to leave room for the possibility of a creature that does not attain the full glory of conceptualised mentation, yet which enjoys conscious experiences with non-conceptual content - experiences that play a role in the explanation of the creature's spatial behaviour.

Despite this difference from Evans over the notion of conscious experience, however, the key idea about perceptual content remains (1982, p. 159):

> It is not necessary, for example, that the subject possess the egocentric *concept* 'to the right' if he is to be able to have the experience of a sound as being to the right. I am not requiring that the content of conscious experience itself be conceptual content.

With so much by way of clarification of the notion of perceptual content, we can now turn to a second preliminary matter: what is needed to establish externalism.

III. THE EXTERNALIST CLAIM

My aim is to establish a *modal* externalist claim. I shall introduce that claim by contrasting externalist claims with corresponding *individualist* claims, and by drawing a distinction between modal and *constitutive* claims in each case.

Here, to begin with, is a statement of *constitutive individualism* (Burge, 1986, pp. 3-4):

> Individualism is a view about how kinds are correctly individuated, how their natures are fixed. . . . According to individualism about the mind, the mental natures of all a person's or animal's mental states (and events) are such that there is no necessary or deep individuative relation between

> the individual's being in states of those kinds and the
> individual's physical or social environments.

I take this to mean that the most fundamental philosophical account of what
it is for a person or animal to be in the mental states in question does not
need to advert to that individual's physical or social environment, but only to
what is going on within the spatial and temporal boundaries of the creature.
     Suppose for a moment that that were right - that constitutive
individualism were correct - and imagine that some individual x is in some
mental state S. Imagine, too, that y is a duplicate of x in the same, or in
another, possible situation. Then, the constitutive account of what it is for x
to be in mental state S will be satisfied equally by y, since that account
adverts only to features that x and y have in common as duplicates. So, if
the constitutive individualist claim were correct for mental state S, then that
state would be preserved across duplicates, whether in the same, or in
different, possible situations.
     In short, the constitutive individualist claim about a family of
mental states or properties entails *modal individualist* claims about those
states or properties - claims to the effect (Burge, 1986, p. 4) that they

> could not be different from what they are, given the
> individual's physical, chemical, neural, or functional
> histories, where these histories are specified non-
> intentionally and in a way that is independent of physical or
> social conditions outside the individual's body.

Such a modal individualist claim is a claim about supervenience: the mental
states or properties in question supervene upon physical, chemical, neural, or
functional states or properties. More specifically, it is a claim of *local
supervenience*, since it says that the mental states or properties of an
individual are fixed by what goes on - physically, chemically, neurally, or
functionally - within the boundaries of that individual's body. If a mental
state or property of an individual x is locally supervenient, then any other
individual y that is a duplicate of x (is the same from the skin inwards) shares
that state or property.
     I speak of modal individualist claims in the plural, because
supervenience claims vary in strength along modal dimensions. What is
principally at issue here is a *modally strong* local supervenience claim of the
form: If x has mental property F in possible world w_1, and y is a duplicate

in w_2 of x (in w_1), then y has F in w_2. It is important to distinguish this
modally strong claim from a claim that is restricted to counterfactual
duplicates of *actual* individuals, and also from a claim that concerns only
duplicates within the *same* possible world: If x has intentional property F in
possible world w, and y is a duplicate in w of x, then y has F in w. (For a
taxonomy of supervenience claims, and a map of their entailment relations,
see McFetridge, 1985. In his notation, the modally strong claim concerns
(XYWW´) supervenience; and the claim about counterfactual duplicates of
actual individuals concerns (XYAW) supervenience. The claim about
duplicates within the same possible world concerns (XYWW) supervenience.)

Just as the various modal individualist claims are entailed by
constitutive individualism, so also the negations of those modal claims entail
the following statement of *constitutive externalism*:

> According to *externalism* about the mind, the mental natures
> of at least some of a person's or animal's mental states (and
> events) are such that *there is a* necessary or deep
> individuative relation between the individual's being in
> states of those kinds and the individual's physical or social
> environments.

I take this to mean that the most fundamental philosophical account of what
it is for a person or animal to be in the mental states in question does advert
to that individual's physical or social environment, and not only to what is
going on within the spatial and temporal boundaries of the creature.

Constitutive externalism is entailed by the negation of the modally
strong local supervenience claim; and it is this relatively modest modal
externalist claim for which I shall be arguing in the case of perceptual
content. My aim is to provide an example in which duplicates x and y,
embedded in possible circumstances w_1 and w_2 respectively, differ in respect
of the non-conceptual contents of their perceptual experiences.

IV. THE TASK FOR AN EXTERNALIST ARGUMENT

Because the conceptual contents of judgements and beliefs are different in
kind from the perceptual contents of experiences, externalist arguments about
the one cannot necessarily be used to defend externalist claims about the
other. In this section, I shall outline the task that confronts the externalist

about perceptual content. The three main points that need to be made correspond to the three features of perceptual content that we noted in Section 2: it is non-conceptual, it is not object-involving, and it is fully representational.

First, some celebrated externalist arguments about belief content (e.g. Burge, 1979) are designed to show that the contents of certain beliefs depend in part upon the *social* context of the believer. In these arguments, a modal externalist claim is defended by way of examples in which the social environment differs as between the possible worlds w_1 and w_2.

Such social externalist arguments about belief content often seem to depend upon the social character of public language meaning - a dependence mediated by a presumed close tie between, on the one hand, belief content itself and, on the other hand, the linguistic meaning of expressions and reports of belief. But, because perceptual content is non-conceptual content, it is not so plausible that it is dependent upon the subject's membership in a speech community. Indeed, perceptual content is reasonably assumed to be independent of public language (cf. Burge, 1986, p. 26). So, we shall not expect to find social externalist arguments about perceptual content.

This independence of perceptual content from linguistic meaning has other consequences, too. Many familiar 'Twin Earth' arguments for externalism in the case of belief content - environmental as well as social - go in step with arguments for the externalism of meaning. Indeed, the line of argument began with meaning (Putnam, 1975), and was then transposed to belief (see McGinn, 1989, p. 31). In the case of perceptual content, a different argumentative strategy is required.

Of course, in the case of belief content - particularly the contents of *de re* beliefs - there are environmental externalist arguments that proceed directly, rather than via externalism about meaning. But - this is the second point that needs to be made - because perceptual content is not object-involving, externalism about perceptual content cannot be established by arguments analogous to externalist arguments about *de re* beliefs.

If I look at an apple, Fido, and think, 'That apple is rotten', and you look at a numerically distinct but qualitatively indistinguishable apple, Fifi, and think, 'That apple is rotten', then - be we ever so similar internally - our beliefs have different contents in virtue of our being related to different apples. My belief, concerning Fido, to the effect that it is rotten, is a belief whose correctness depends upon how things are with Fido: whether Fido is indeed a rotten apple. Your belief, in contrast, is one whose correctness is indifferent to how things are with Fido, but depends instead upon how things are with Fifi. In that sense, the contents of our beliefs are object-involving. As a result, it is easy to generate a modal externalist example just by varying the object of belief as between the possible worlds w_1 and w_2. But this strategy is not available in the case of perceptual content.

We introduced the idea that perceptual content is not object-involving in the context of Peacocke's account of scenario content in terms of ways of filling out the space around the subject. We can now connect the idea with the thought that the perceptual content of experience is a phenomenological notion: perceptual content is a matter of how the world *seems* to the experiencer (Evans, 1982, p. 154; McGinn, 1989, p. 66). If perceptual content is, in this sense, 'phenomenological content' (McGinn, *ibid.*) then, where there is no phenomenological difference for the subject, there is no difference in perceptual content. So, if two objects are genuinely indistinguishable for a subject, then a perceptual experience of the one has the same content as a perceptual experience of the other. This is in sharp contrast to the case of belief content, since the intuition about the content of *de re* beliefs concerning Fido and Fifi carries over to the case of two beliefs held by a single subject.

While perceptual content is not object-involving, it is still fully representational: the content of a perceptual experience determines a condition for correctness, or truth. One way to see how perceptual content can be truth conditional, although not object-involving, is to take perceptual content to be existentially quantified content. A visual experience may present the world as containing *an* object of a certain size and shape, lying at a certain distance from the subject, in a certain direction. It matters not at all to that existentially quantified content of a subject's experience whether, for example, it is Fido or Fifi that she is looking at.

The third point that needs to be made is that, because perceptual content is fully representational although not object-involving, the individualist about perceptual content is in a very different position from the individualist about belief content.

In the case of object-involving belief content, there are familiar proposals to factor the content into two components. There is one component that the content of my belief about Fido has in common with the content of your belief about Fifi; and there is another component that is not shared - a component that determines the involvement of the particular apple Fido in the correctness conditions of my belief (e.g. McGinn, 1982). The first component - the *narrow* content of the belief - is supposed to be locally supervenient, and so preserved across actual and counterfactual duplicates.

But it does not, by itself, determine truth conditions, since what is in common between your belief and mine does not, by itself, specify whether the correctness of my belief turns upon how things are with Fido or upon how things are with Fifi. The second component is a matter of how things are in my environment. More specifically, the second component concerns causal relations between my brain and a particular object in my environment, namely Fido.

In the context of a dual component, or two factor, proposal of this kind, the individualist typically concedes an externalist claim for the truth conditional content of beliefs, but says that, for serious explanatory purposes, attention should be restricted to narrow content. Thus, the individualist about belief content recommends the employment of a kind of content that is locally supervenient, but is not fully representational (Fodor, 1986; 1987, Chapter 2).

Because of the differences between perceptual content and belief content - especially, because perceptual content is not object-involving - it is open to the individualist about perceptual content, in contrast, to say that experiences have content that is both locally supervenient and fully representational. Indeed, I shall take it that the individualist makes just this bolder claim, and that this is what the externalist has to argue against.

In order to establish his case, the externalist is obliged to produce a persuasive example with the following structure. First, in some possible situation w_1 - perhaps the actual situation - a subject x has an experience with a certain existentially quantified content. For example, it might be an experience as of a square object of a certain size (cf. McGinn, 1989), or an experience as of a shadow of a certain size and shape (cf. Burge 1986, 1988a). Second, a duplicate subject y in some other possible situation w_2 has an experience which, despite everything being the same from the skin inwards, does not have that same content. This is all that is required to refute the modally strong claim of local supervenience, and establish the modest modal externalist claim.

But the externalist may choose to go further by trying to make it plausible, not merely that the duplicate's experience does not have the same content as the original subject's experience, but also that the duplicate's experience has some specific alternative content. It might be that the duplicate's experience is as of a round object, instead of as of a square object, or that the duplicate's experience is as of a crack, instead of as of a shadow.

V. TWO INDIVIDUALIST STANCES AND A DILEMMA FOR THE EXTERNALIST

Given a putative externalist example with this structure, the individualist may adopt one of two possible stances. The individualist who adopts a *conservative* stance towards an example accepts the externalist's specification of the content of the experience in the original possible situation (say, the actual situation). But the individualist then rejects the externalist's claim that the experience of the duplicate in the alternative possible situation does not have that same content. Thus, for example, an individualist adopting a conservative stance may accept that an actual subject has an experience as of a shadow; but the individualist then insists that the duplicate subject's experience is also as of a shadow, despite the environmental differences.

The individualist who adopts a *revisionary* stance towards an example does not accept the externalist's specification of the content of the experience in the original possible situation. Thus, for example, an individualist adopting a revisionary stance might agree that, *if* an actual subject's experience is as of a shadow, then the experience of a duplicate may differ in content. But the individualist insists that the specification of the content of the actual subject's experience - as of a shadow - is unmotivated. The experiences of both the actual subject and the duplicate subject should be assigned some more inclusive content - perhaps: as of a shadow-or-crack.

Robert Matthews illustrates how to adopt each kind of stance in response to versions of Tyler Burge's (1986, 1988a) example of the shadows and cracks. In Burge's story, an individual P normally perceives Os (shadows of a certain small size) as Os, but occasionally misperceives a C (a similarly sized crack) as an O. In a counterfactual situation (1988a, pp. 75-6):

> there are no visible Os . . . [and] . . . the visual impressions caused by and explained in terms of Os in the actual situation are counterfactually caused by and explained in terms of Cs - relevantly sized cracks. The cracks are where the shadows were in the actual case.

In the actual situation, the subject P sees shadows as shadows and occasionally sees a crack as a shadow. Concerning the counterfactual situation, Burge makes a bolder and a more cautious claim. The bolder claim is that, 'Counterfactually, . . . P sees Cs as Cs' (1988a, p. 76); that is, the duplicate sees cracks as cracks. The more cautious claim is just that the duplicate does not see the cracks as shadows (1988b, p. 95):

> [N]othing in the argument depends on attributing any
> specific perceptual states to the organism in the
> counterfactual situation. All that is important is that it be
> plausible that the counterfactual perceptual states are
> different from those in the actual situation. So the question
> about whether . . . the organism perceives cracks as cracks
> in the counterfactual situation is not directly relevant to the
> argument.

This latter claim reflects just how little is dialectically required of the externalist. He only has to make it plausible that the subject in the counterfactual situation differs from the subject in the actual situation to the extent of not seeing the cracks as shadows.

Matthews demonstrates a conservative individualist stance as follows. In Burge's illustration, 'we may imagine that the sort of entities being perceived are very small and are not such as to bear on the individual's success in adapting to the environment' (1988a, p. 75). But suppose instead, says Matthews (1988, p. 83):

> that the shadows and cracks in question are important to the
> organism's adaptive success, e.g., that the shadows are
> important sources of shade for the organism during the heat
> of the day, and that the cracks are large enough that the
> organism risks injury if it should fall into them.

In the actual situation, then, the organism will go towards whatever is seen as a shadow, and avoid whatever is seen as a crack; the type of experience that is normally produced by shadows will be connected to dispositions to produce certain bodily movements. It is built into Burge's example that behavioural dispositions are the same in the counterfactual situation as in the actual situation. So, likewise in Matthews's variant, the same connections to bodily movements will be present in the duplicate in the counterfactual situation in which there are cracks where the shadows were.

Now, the externalist is supposed to make it plausible that the duplicate sees cracks as cracks, or at least not as shadows. But (Matthews, 1988, p. 83):

> If in the counterfactual environment the organism repeatedly
> fell into the cracks when during the heat of the day it sought
> shelter from the sun, we would surely conclude that in this
> environment the organism perceives cracks as shadows, or
> at least not as cracks.

The behaviour consequent upon the organism's visual experiences in the
counterfactual situation supports the attribution of the very same content as
in the actual situation. (If the individualist is adopting a conservative stance
then it is not adequate merely to argue that the organism does not see cracks
as cracks; he must maintain that the organism sees cracks as shadows.)

Matthews shows how to adopt a revisionary stance in response to
Burge's original version of his illustration in which no particularly adaptive
behaviour is produced as a result of the type of experience that is normally
caused in the actual situation by shadows. In this case (Matthews, 1988, p.
83):

> Burge has provided no reason for supposing that in the
> counterfactual environment the organism perceives cracks as
> cracks. Of course, there is no reason to suppose that in the
> counterfactual environment the organism perceives cracks as
> shadows, but it hardly follows from this that it perceives
> cracks as cracks. Given that the organism does not
> discriminate cracks from shadows . . . one could as well
> argue that this organism perceives cracks and shadows as
> instances of one and the same type of entity.
> An organism may perceive O's in the actual environment and
> C's in the counterfactual environment, not as O's or C's, but
> rather as instances of an objective type that includes both
> O's and C's.

If the behaviour that is consequent upon a type of experience is equally
appropriate to a shadow and to a crack then we have no compelling reason to
say that the experience is as of a shadow or that it is as of a crack. Just as
we now lack a reason to say that the duplicate's experience is as of a shadow,
so also we lack a reason for claiming that the experience of the organism in
the actual situation is as of a shadow. Rather, we should say that both actual
and counterfactual experiences are as of a shadow-or-crack. Thus the
revisionary individualist stance. (For a detailed development of this
revisionary stance, see Segal, 1989. For a rejection of adopting the stance

towards Burge's example, see Davies, 1991; and for a rejoinder, see Segal, 1991.)

What lesson should we draw from the individualist's adoption of one or the other of these stances towards the externalist's examples? One lesson concerns causal covariance theories of perceptual content.

Externalism is easy to establish if we take as a premise a covariance theory of content. For, according to such a theory, if the (predominant) causal antecedents of a type of experience are changed as between the actual and counterfactual situations, then the content of experiences of that type is changed, too.

But, causal covariance theories do not merely entail externalism. Covariance theories are pure input-side theories that nowhere advert to output factors such as behaviour. Consequently, they impose constraints upon putative externalist examples; particularly, upon pairs of examples that differ only in the behavioural consequences of experiences, and not in the causal antecedents of experiences. If two examples differ in that way, then they should agree in the content they assign to the organism's actual experience and in the content they assign to the duplicate's experience in the counterfactual situation.

Matthews's adoption of a conservative individualist stance exploits this consequence, and thereby casts doubt upon covariance theories. For Matthews's variant of the example of the shadows and cracks differs from Burge's own version of his illustration only in the causal consequences of experiences. Yet, it is markedly less plausible to say that the duplicate sees cracks as cracks in Matthews's variant than it is in Burge's original version.

This certainly counts against causal covariance theories of perceptual content. But it does not count straightforwardly against externalism unless the externalist is committed to a covariance theory; that is, unless the externalist is committed to saying that a difference in causal antecedents is sufficient for a difference in content. Is the externalist so committed?

Matthews seems to see such a commitment in Burge: 'we consider a modification of Burge's example that *should*, but does not, *leave his conclusion intact*' (1988, p. 82; emphasis added). But Burge himself stresses that he is not committed to any sufficient condition for an experience to have a particular content (1988b, p. 93); and, so far from opting for a covariance theory, he regards evolutionary factors, for example, as relevant to the attribution of perceptual content (e.g. 1986, p. 40). In any case, it is clear that a causal covariance theory of content is not an essential requirement in an externalist argument. For, to rebut the modally strong claim of local supervenience, all that the externalist has to show is that there are some environmental differences between situations w_1 and w_2 - however thoroughgoing - that suffice for a difference of perceptual content between duplicates x in w_1 and y in w_2.

Objections to covariance theories of content are not automatically objections to externalism. But still, the inadequacy of covariance theories serves to highlight the fact that attributions of perceptual content - particularly, contents involving shape and distance properties - are partly answerable to the subject's behaviour; and this fact presents the externalist with something of a dilemma.

For, either the subject in the actual situation produces behaviour that is particularly appropriate to the supposed content of her experience, or else she does not. If she does, and that behaviour perseveres into the counterfactual situation, then the individualist may adopt a conservative stance, insisting that the duplicate subject's experience has that same content. If she does not, then the individualist may adopt a revisionary stance, maintaining that the specification of content for the actual subject's experience is unmotivated.

This dilemma is particularly pressing for the externalist who sets out to show just what Burge aims to show with his example of the shadows and the cracks; namely, that perceptual content does not supervene upon internal constitution *plus behavioural dispositions* (1986, p. 39; 1988a, p. 69). But it is important to notice that this is strictly speaking more than the externalist is obliged to demonstrate. The externalist is allowed to have the duplicate's behavioural dispositions differ from those of the actual subject, to the extent that this is consistent with the two being duplicates.

This may appear to be a negligible degree of freedom for the externalist since, surely, the basis of behavioural dispositions is to be found inside the skin. But, if behaviour is itself characterised externalistically, then the production of behaviour of a certain type depends both upon what happens inside the skin - nerve firings, muscle contractions, and the like - and upon environmental factors. In principle, behaviour - externalistically characterised - can be varied even while everything inside the skin remains the same.

The externalist carries the day if, taking advantage of this freedom, he can construct a persuasive example against which neither a conservative nor a revisionary individualist stance can plausibly be adopted.

VI. EXTERNALISM VINDICATED

In this section, my aim is to provide - at least in outline - a persuasive externalist example. I shall first present the example in schematic form, and then sketch an instantiation of the schema by giving a twist to an example of McGinn's (1989, pp. 63-8).

6.1 *A Schematic Example*

First, in some possible situation w_1 - let us say, the actual situation - a subject x enjoys experiences with perceptual content. On the input side, perceptual states of intrinsic type T covary with the distal occurrence of visual property O. (We might think of O as a shape property, say, being square, or a distance property, say, being one metre away, or a direction property, say, being 20 degrees to the left of straight ahead.) On the output side, the behaviour of type B that is consequent upon internal states of type T is particularly appropriate to O's occurrence. Thus, the input side - distal antecedents - and the output side - behavioural consequences - are in harmony; and we can suppose that evolutionarily this is no accident. We may assume that in securing covariation between T and O, the subject's visual system is doing just what it is supposed to do. Aspects or components of the visual system have been selected for their having the consequence that internal states of type T covary with occurrences of property O.

Even without a detailed theory of perceptual content, it does not seem illegitimate to suppose that, by elaborating these input-side, output-side, and teleological factors, we can make it plausible that the subject sees Os as Os; that is, that in the actual situation the perceptual states of type T are experiences as of an O.

Second, in some other possible situation w_2 - a counterfactual situation - there is a duplicate y of x. This counterfactual situation is different from the actual situation in respect of the environment and perhaps also the laws of nature. As a result of these differences, distal occurrences of visual property C produce just the same retinal array as do occurrences of O in the actual situation. Consequently, perceptual states of the intrinsic type T covary with the occurrence of C, rather than of O. (We might think of C as a different shape property, say, being round, or a different distance property, say, being 75 cm away, or a different direction property, say, being 30 degrees to the left of straight ahead.)

Because y has the same internal constitution as x, states of type T
have just the same internal consequences, such as nerve firings and muscle
contractions, as in the actual situation. But, environmental differences in - as
it might be - gravity or friction conspire to produce trajectories for y's body
that are quite different from those carved out by x's body in the actual
situation. Thus, input-output harmony is preserved: the behaviour of type D
that is counterfactually consequent upon internal states of type T is
distinctively appropriate to C's occurrence, rather than to O's. Furthermore,
y's visual system is doing just what it is supposed to do. The ancestors of y
have led full and happy lives and had lots of babies in part because internal
states of type T covary with occurrences of C.

Once again, even without a constitutive theory of perceptual content
to hand, it seems reasonable to suppose that we can make it plausible that
the duplicate subject sees Cs as Cs; that is, that in the counterfactual
situation the states of type T are experiences as of a C. *A fortiori*, we can
make it plausible that those states are not experiences as of an O; and this
latter claim is all that is needed to rebut local supervenience.

An example of this form cannot, of course, be used to demonstrate
that perceptual content fails to supervene on internal constitution *plus*
behavioural dispositions. For although we can plausibly vary perceptual
content as between the actual and the counterfactual situation, we also vary
behavioural dispositions if these are externalistically characterised in terms of
bodily trajectories. But, just as it stands, a persuasive example of this form
carries the day against individualism. For it presents a difference of
perceptual content between duplicates; and that is enough to establish all that
the externalist is dialectically obliged to establish, namely that (Burge, 1986,
p. 4):

> A person's intentional states and events could
> (counterfactually) vary, even as the individual's physical,
> functional (and perhaps phenomenological) history . . . is
> held constant.

Whether it is possible to modify such an example so as to vary perceptual
content while keeping the behavioural dispositions the same is a subsidiary
question that is not my main concern here.

6.2 *Percy*

The example to which I shall, in the next sub-section, give a twist is actually used by McGinn (1989, pp. 58-94) to argue *against* what he calls *strong externalism* for perceptual content. McGinn's target is the thesis that the difference between an experience of something looking square and an experience of something looking round is 'a matter of a difference in how those experiences relate to instantiations of squareness and roundness' (1989, p. 63). In essence, what McGinn aims to rebut is a causal covariance theory of perceptual content.

     To this end, McGinn constructs an example. In the actual situation, internal state S1 of the subject Percy is caused by square things and internal state S2 is caused by round things. In the counterfactual situation, Percy's internal constitution and behavioural dispositions are just as in the actual situation, but as a result of environmental differences, state S1 is produced by round things and state S2 is produced by square things. On a particular occasion in the counterfactual situation, Percy is in state S1. Is the perceptual content of his experience that there is a square thing before him or that there is a round thing before him? Is the experience as of something square or as of something round?

     The strong externalist must say that the content of the experience in the imagined case is individuated in terms of the distal causes of state S1 in the counterfactual situation; thus the content of Percy's experience is that there is a round thing before him. McGinn, in contrast, argues that in the counterfactual situation Percy is doomed to misperceive round things as square. In support of this view, McGinn points to the fact that Percy's behaviour, consequent upon his being in internal state S1, is appropriate to the presence of a square thing - for behavioural dispositions are preserved across the actual and counterfactual situations. He makes it plausible that, where there is dislocation between the facts of covariance on the input side and the facts of behaviour on the output side, output-side factors should dominate in the ascription of perceptual content (1989, p. 66): 'So when it comes to a competition between action and environment, in the fixation of perceptual content, action wins.'

     Furthermore, McGinn points out, this judgement about the content of Percy's experiences is backed up by teleological elements that plausibly belong in a theory of perceptual content (1989, pp. 66-7):

> We naturally want to say that the *purpose* of his moving in a

> square path is to negotiate square objects successfully, that
> this is the *function* of his moving like that.
>   . . . if Percy's functional properties are preserved [in the
> counterfactual situation], so too will be the content of . . .
> his perceptual states. That is, if his squarewise movements
> have the function precisely of negotiating square things,
> then the perceptual states that lead to these movements will
> partake of this function and have their contents fixed
> accordingly.

We shall surely agree with McGinn in rejecting strong externalism here. In effect, he is adopting a conservative individualist stance towards a particular example; and his attitude towards Percy in the counterfactual situation is much like Matthews's attitude towards the creatures who keep falling into the cracks.

But, McGinn's argument does not, of course, show that individualism is correct for perceptual content; it does not establish a modally strong claim of local supervenience. Nor does McGinn set out to defend individualism, as we have defined that doctrine. So, there is nothing inconsistent in adapting McGinn's example in the service of externalism.

In fact, the issue of externalism - in the sense of the denial of local supervenience - lies somewhat off to one side from McGinn's main concerns. For, in his thought experiments, McGinn includes behavioural dispositions among the internal factors that are to be held constant across actual and counterfactual scenarios (1989, p. 2). But, as we have seen, if behavioural dispositions are characterised externalistically in terms of bodily trajectories, then they can vary even while all that is inside the skin stays the same. This point will be crucial when we come to give a twist to the example of Percy.

Before we move on, however, it is tempting to pause briefly and ask whether the example of Percy makes it plausible that perceptual content supervenes upon internal constitution *plus behavioural dispositions*. There are grounds for supposing that it does not. In the counterfactual situation we have Percy (or a duplicate) moving squarewise in response to round things; and McGinn makes it plausible that the character of the behaviour is more important for perceptual content than are the distal antecedents. This intuition is particularly strong when the function of the behaviour is preserved along with its spatial character. Nevertheless, it does seem possible that, where there is a sufficiently hopeless breakdown of harmony between input-side and output-side factors, we may be entitled to withhold all attributions of perceptual content (Fricker, 1991, p. 141). Consequently, there could be an example of duplicates, sharing their behavioural dispositions, yet differing in that one has experiences with perceptual content and the other does not.

6.3 *Percy with a Twist*

Let us slightly vary McGinn's example. In the counterfactual situation we now find, not Percy himself, but a duplicate with a very different evolutionary history. This creature's ancestors survived to reproduce in part because their behaviour was appropriate to the distal causes of their perceptual experiences. In this imaginary scenario, internal state S1 is produced by distal round things, as in McGinn's example; but the behaviour consequent upon the creature's being in S1 is now appropriate to the presence of round things, and not to the presence of square things.

What is being imagined here is not that walking a square trajectory is the best way of avoiding a round object. Rather, we suppose that environmental differences have the consequence that the same nerve firings and muscle contractions as in the actual situation result in a quite different bodily trajectory. In particular, the goings-on inside the skin which in the actual situation lead to a square trajectory now have a round trajectory as their upshot. This happy agreement of input-side, output-side, and teleological factors makes it plausible that, when Percy's duplicate is in state S1, he has an experience as of a round thing. *A fortiori*, it is implausible that the duplicate misperceives round things as square; and this is all that the externalist argument strictly requires.

What we have here is, of course, just an instantiation of our schematic example, with the shape properties of being square and being round now playing the roles of O and C. And what goes for shape properties surely goes equally - or even more so - for distance and direction properties. But, perhaps some individualist critic will deny that this is a persuasive externalist example, on the grounds that the departures from actuality required by the substitution of circles for squares are wildly science fictional.

It is unclear that this is an effective individualist response, since the whole discussion has been carried out in the domain of thought experiments; and, in the face of the individualist's modally strong claim of local supervenience, it is surely legitimate to consider counterfactual situations that are also counternomic. Certainly Burge is explicit that, 'Since examples usually involve shifts in optical laws, they are hard to fill out in great detail'

(1986, p. 42). But, perhaps we can do something to reduce the wildness.

Instead of considering squares in the actual situation, let us consider ellipses. In particular, let us consider ellipses that are slightly elongated along the (gravitational) vertical axis. Our perceiver Percy sees these ellipses as ellipses - as witness input-side, output-side, and teleological factors surrounding his internal state S1. In addition, in the actual situation, Percy sees round things as round (and is then in internal state S2).

In the counterfactual situation, we imagine that the retinal arrays - and consequent internal state S1 - that are actually produced by these vertically elongated ellipses are instead produced by circles; and behaviour is squashed along the vertical axis so that input-output harmony is preserved. Furthermore, we imagine all this to be the result of evolution. Percy's duplicate is as well adapted to this counterfactual situation as Percy is to the actual situation.

The plausible externalist claim about this example is that, when Percy's duplicate is in the same internal state S1 that Percy is in when he has an experience as of a vertically elongated ellipse, the duplicate's experience is as of a round thing. *A fortiori*, the duplicate's experience is not as of an ellipse; and this is all that the externalist argument strictly requires.

It is as well to enter two clarificatory comments about this example. The first concerns axes of symmetry. Since an ellipse has just two axes of symmetry while a circle has infinitely many, someone might ask how many axes of symmetry Percy's duplicate sees circles as having. This is a question about the representational properties of the duplicate's perceptual experiences. (It is a matter of protopropositional content: Peacocke, 1992, p. 77.) So, it would be begging the question against externalism if someone were to insist that the duplicate must see circles as having just two axes of symmetry. Nevertheless, it is open to the externalist to allow that Percy's duplicate sees circles as having only vertical and horizontal axes of symmetry, just as Percy sees ellipses as having only two axes of symmetry. For it is not uncommon for subjects to see shapes as having fewer axes of symmetry than they really have. For example, a square has four axes of symmetry, but when seen as a square it is seen as having two axes of symmetry (intersecting the sides), and when seen as a diamond it is seen as having a different two (intersecting the corners).

The second clarificatory comment concerns the simplifying assumptions that are implicit in the example. The duplicate's behaviour is supposed to be squashed along the (gravitational) vertical axis, in virtue of some environmental difference in, say, gravity or friction. But, there is an implicit assumption here, to the effect that the range of Percy's behavioural interactions with ellipses is very limited. This does not undermine the dialectical purpose of the example of Percy and his duplicate. But it does suggest that it will be difficult to provide an externalist example relating to the experiences of shape that are enjoyed by creatures whose interactions with the world are as complex and sophisticated as ours are (see Davies, 1993).

So much, then, for what I claim to be (a sketch of) a simple but persuasive externalist example involving shape properties. (It is a straightforward matter to produce similar examples involving distance or direction properties.) Can either a conservative or a revisionary individualist stance be adopted towards the example of Percy and his duplicate? Neither looks plausible.

The individualist who adopts a conservative stance towards the example accepts the externalist's specification of the content of the experience that Percy enjoys when he is in internal state S1. It is an experience as of a vertically elongated ellipse (or as of a square, in the first version of the example). But the individualist then insists that the experience of Percy's duplicate has that same content; that the duplicate misperceives round things as elliptical, despite producing behaviour that is distinctively appropriate to the occurrence of roundness. Given the convergence of input-side, output-side, and teleological factors in the example, the conservative individualist stance appears quite unmotivated.

But a revisionary individualist stance looks even less attractive. To adopt this stance is to deny that Percy's actual experience is as of an ellipse, and to say that the experiences of both Percy - in the actual situation - and his duplicate - in the counterfactual situation - should be assigned some more inclusive content, such as: as of an ellipse-or-circle. But, if that is the content of Percy's actual experience when he is in state S1, then why does he execute behaviour that is particularly appropriate to vertically elongated ellipses? And what is the content of his experience when he is in state S2 (produced by round things)? In short, the adoption of a revisionary individualist stance is problematic for the intentional explanation of Percy's behaviour (see Davies, 1991).

Thus is externalism concerning perceptual content vindicated. But the vindication leaves us with a puzzle about the phenomenology of perceptual experience. This puzzle is the topic of my final section.

VII. PERCEPTUAL CONTENT AND PHENOMENOLOGY

Our externalist conclusion that perceptual content is not locally supervenient - that it does not strongly modally supervene upon the internal state of the subject - appears to be inconsistent with the conjunction of two antecedently plausible propositions about phenomenology.

The first of these two propositions is that perceptual content is a matter of how things seem to the conscious subject. Thus, for example, McGinn insists (1989, p. 63):

> Let us be clear that we are considering a phenomenological notion here: conscious seemings, states there is something it is like to have. . . .
> So we are considering properties of organisms that determine the form of their subjectivity, . . .

Perceptual content is a matter of how the world is presented to the conscious subject as being arranged.

The second proposition is that experience has a phenomenal character that is supervenient upon the internal state of the subject. The intuition here is that neither the nature of the distal antecedents, nor the shape of the consequent trajectory, nor the course of evolutionary history, is a determinant of the subjective character - the 'what it is like' - of sensory experience. According to this second proposition, what it is like, phenomenally, to be Percy is just the same as what it would be like to be Percy's duplicate.

If perceptual content is a phenomenological notion - as the first proposition says - then the inescapable conclusion appears to be that perceptual content supervenes upon whatever the phenomenal character of the subject's experience supervenes upon. But then, by the second proposition, perceptual content turns out to supervene upon internal constitution - in contradiction with our externalist conclusion.

The first proposition says that the way the world is presented to the subject is a matter of phenomenology; perceptual content supervenes on phenomenal character. The second proposition says that phenomenology is locally supervenient: phenomenal character supervenes upon internal constitution. The clash with externalism then appears to be a consequence of the transitivity of supervenience.

But, we can see our way to one possible resolution of the puzzle if we are more careful about the notion of supervenience that is at work in the first proposition about phenomenology.

7.1 A Resolution: Kinds of Supervenience

In Section 3, we noted that supervenience claims vary in strength along modal dimensions, and we distinguished 'across worlds' (XYWW′) supervenience - which has been our principal concern in this paper - from, for example, 'within a world' (XYWW) supervenience. So, the question to ask is: What kind of supervenience is involved in the first proposition about phenomenology (perceptual content supervenes upon phenomenal character)?

We are certainly committed to the claim that perceptual content is a phenomenological notion. Indeed, we have already used that claim (in Section 4) to support the idea that perceptual content is not object-involving. As McGinn says (1989, p. 63), 'Looking square is subjectively distinct from looking round': where there is a difference of perceptual content, there must be some difference in the phenomenal character of the experiences. Supervenience is surely in the offing here. But, in order to honour the phenomenologicality of perceptual content, we only need this supervenience to apply within an individual subject, in a single possible world. If there is no phenomenal difference between two experiences in the life of a given subject, then those experiences have the same perceptual content.

From this 'within a subject, within a world' (XXWW) supervenience claim - however modally strong may be the supervenience claim in the second proposition about phenomenology - no amount of transitivity will take us to the denial of our externalist conclusion. In short, the apparent puzzle is generated by a failure to distinguish between the modally modest supervenience that is used in the first proposition about phenomenology and the modally strong supervenience that is at issue in the debate between individualism and externalism.

The distinction between kinds of supervenience claim yields a resolution of the puzzle about externalism and phenomenology by trading upon another distinction; namely, that between the perceptual content of an experience and its intrinsic phenomenal character. Perceptual content is a matter of representational properties of experience, while intrinsic phenomenal character is conceived of as being non-representational.

Given the distinction between perceptual content and phenomenal character, the externalist about perceptual content can accept both the

propositions about phenomenology. He accepts the first proposition by
saying that, for the experiences of a given subject, a difference of perceptual
content requires a difference in intrinsic phenomenal character. Within a
subject, and within a world, a representational difference requires a non-
representational difference. He accepts the second proposition by saying that
the (non-representational) intrinsic phenomenal character of an experience
really is strongly modally locally supervenient: it is preserved across actual
and counterfactual duplicates.

His externalism then commits him to the possibility of a difference
of perceptual content between the experiences of (actual and counterfactual)
duplicates; that is - by the second proposition - to the possibility of a
difference of perceptual content even while intrinsic phenomenal character is
preserved. Across subjects, and across worlds, a representational difference
does not require a non-representational difference. Indeed, this possibility is
explicitly recognised in Burge's expression of modal externalism (1986, p. 4;
emphasis added):

> A person's intentional states and events could
> (counterfactually) vary, even as the individual's physical,
> functional (*and perhaps phenomenological*) history . . . is
> held constant.

But, before we rest content with this resolution of the puzzle, we should ask
ourselves whether we want to take on this commitment to the intrinsic
phenomenal character of experience.

7.2 Sensational Properties of Experience

The idea of a level of intrinsic phenomenal character, intermediate between
internal physical constitution and perceptual content, is the idea of perceptual
experience as having a sensational (non-representational) substrate upon
which the representational superstructure of perceptual content is erected.
Appealing to this intermediate level so as to resolve the puzzle about
externalism and phenomenology commits us to there being a sensational
difference corresponding to every representational difference within the
experience of a given subject. That commitment goes well beyond the mere
acknowledgement that perceptual experiences have sensational as well as
representational properties.

Certainly some who reject the idea of a sensational substrate take the further step of rejecting sensational properties altogether. Thus, McGinn points out (1989, p. 75) 'obscurities and problems' that beset the view that recognises (1989, p. 73):

> a prerepresentational yet intrinsic level of description of experiences: that is, a level of description that is phenomenal yet noncontentful . . .

and accompanies this with the strong claim that 'perceptual experience has none but representational properties (at least so far as consciousness is concerned)' (1989, p. 75). But, it is surely an option to acknowledge the existence of sensational properties of experience without embracing the idea of a sensational substrate.

The distinction between representational and sensational (intrinsic but not representational) properties of experience is the focus of earlier work by Peacocke (1983, Chapter 1). There, he offers examples that are intended to show that there could be pairs of experiences with the same sensational properties but different representational properties, and other examples that are intended to show the converse - that there could be pairs of experiences with the same representational properties but different sensational properties. Now, one of the background assumptions of that earlier work is that perceptual content is conceptual content (1983, p. 19):

> [N]o one can have an experience with a given representational content unless he possesses the concepts from which that content is built up.

As a consequence, many of the lessons drawn from the examples do not carry over into the framework of Peacocke's own later work, and of this paper, where non-conceptual content is recognised. (See Crane, 1992a, for a helpful discussion of these differences.) For example, in the earlier work, grouping phenomena are described in terms of sensational properties (1983, pp. 24-5); in the later work, they are described in terms of protopropositional content (1992, p. 79), which is a kind of non-conceptual content, cutting somewhat more finely than scenario content.

Nevertheless, there is still enough in those examples to make it plausible that perceptual experiences have sensational, as well as representational, properties. In particular, the example of monocular and binocular viewing of the same scene - in a case where the scene provides sufficiently many cues that there is no loss of depth information when only one eye is used (1983, p. 13) - seems to provide a pair of experiences that present the space around the subject as being filled out in just the same way. Yet the two experiences are phenomenologically different. What it is like to have the monocular experience is not just the same as what it is like to have the binocular experience, even though the experiences have the same perceptual content.

Certainly, not everyone is persuaded by this example (e.g. Tye, 1991, p. 130; 1992, p. 174). Further discussion would be warranted. But the important point for present purposes is that acknowledging the existence of sensational properties, on the basis of examples such as this one, is very far from embracing the idea of a sensational substrate. It would take a massive leap to move from a modest non-representational difference between monocular and binocular viewing of the same scene to a host of non-representational properties subvening under the myriad representational properties of every perceptual experience.

7.3 *A Question About Phenomenal Character*

If we do make that leap then, as we have seen, we can resolve the puzzle about externalism and phenomenology. That is, we can accept the modal externalist claim, along with the two plausible propositions about phenomenology. (We might even take the possibility of resolving the puzzle in this way as an argument for making the leap; see Davies, 1992, pp. 42-4.) But, we then face a further problematic question. For, if perceptual content can vary - as between duplicates in different possible worlds - while intrinsic phenomenal character remains the same, then we are bound to ask whether the correlation between sensational and representational properties is relatively constrained or relatively unconstrained. Just how different might be the representational superstructures erected upon one and the same non-representational substrate?

If the correlation between sensational substrate and representational superstructure is relatively constrained, then it must be governed by some theoretical principles. But, it is far from obvious where the required constraining principles might issue from. On the other hand, to the extent that the relation is unconstrained, we are left entertaining scarcely intelligible hypotheses, along the lines that I, or a duplicate, might enjoy an experience with just the same intrinsic phenomenal character as my visual experience now, yet with utterly different representational properties. Neither option is very inviting.

I do not say, definitively, that there is no way to answer this question about the correlation between sensational substrate and representational superstructure. Perhaps some constraining principles will be forthcoming, for example. But, I do say that the question is problematic; that the problem it poses is no less daunting than the original puzzle concerning externalism and the two propositions about phenomenology. Confronted with the problematic question, we should hesitate over a commitment to a sensational substrate - to the intrinsic phenomenal character of experience.

As a result, we should reconsider the second proposition about phenomenology; namely, the proposition that there is a level of phenomenal description of perceptual experience that is (modally strongly) supervenient upon the internal state of the subject. We can accept that perceptual experiences have some sensational properties (on the basis of such examples as monocular and binocular viewing of the same scene); and, indeed, we can suppose that these non-representational properties are strongly modally locally supervenient. But, it is at best an open question whether such sensational properties constitute a subvening basis upon which the representational properties of experience are variously supported in actual and counterfactual situations.

Suppose that we give up the idea of a non-representational underpinning for every representational property of experience. Then we can still make something of the first proposition about phenomenology: perceptual content is a phenomenological notion. We no longer say that perceptual content supervenes (within a subject, within a world) upon something that is supposed to be more fundamentally phenomenal: the non-representational intrinsic phenomenal character of experience. Instead, we say that perceptual content is itself irreducibly an aspect of what it is like to have a perceptual experience. But the second proposition about phenomenology has to be given up. Given externalism, what it is like to have a perceptual experience - now regarded as shot through with perceptual content - is not wholly independent of causal antecedents, consequent trajectory, and evolutionary history. It is not supervenient (across subjects, across worlds) upon the internal state of the subject. If we give up the idea of a sensational substrate, then there is an important difference here between externalist and

individualist conceptions of the subjective nature of sensory experience.
In the first six sections of this paper, I argued in favour of externalism about
perceptual content. Now, in this final section, we see that, once externalism
is accepted, we may well have to give up the idea that experience has any
(strongly modally) locally supervenient level of phenomenal description*.

NOTE

*Acknowledgements: This paper is a descendant of 'Perceptual Content and
Local Supervenience' (Davies, 1992), which appeared in the *Proceedings of
the Aristotelian Society* Volume 92. I am grateful to the Aristotelian
Society for permission to re-use substantial parts of that paper. While many
of the differences are presentational, there has also been a substantive change
in my views about the sensational properties of experiences (Section 3 of
Davies, 1992, and Section 7 of the present paper).

Some of the early work towards the paper was carried out at the
Australian National University in 1990 and at MIT in 1991. I am grateful to
ANU, the British Academy, MIT, and the Radcliffe Trust for financial
support. Thanks to David Bell, Ned Block, Tyler Burge, Michael Glanzberg,
Frank Jackson, Greg McCulloch, Christopher Peacocke, Gabriel Segal, Tom
Stoneham, and Stephen Williams for comments and conversations.

REFERENCES

Block, N.: 1990, Consciousness and Accessibility. *Behavioral and Brain Sciences* vol. 13, pp. 596-8.

Block, N.: 1991, Evidence Against Epiphenomenalism. *Behavioral and Brain Sciences* vol. 14, pp. 670-2.

Block, N.: 1992, Begging the Question Against Phenomenal Consciousness. *Behavioral and Brain Sciences* vol. 15, pp. 205-6.

Block, N.: 1993, Review of D.C. Dennett, *Consciousness Explained. Journal of Philosophy* vol. 90, pp. 181-93.

Block, N.: 1995, On a Confusion about a Function of Consciousness. *Behavioral and Brain Sciences* vol. 18.

Burge, T.: 1979, Individualism and the Mental. In P.A. French, T.E. Uehling and H.K. Wettstein (eds.), *Midwest Studies in Philosophy Volume 4: Studies in Metaphysics*, Minneapolis: University of Minnesota Press, pp. 73-121.

Burge, T.: 1986, Individualism and Psychology. *Philosophical Review* vol. 95, pp. 3-45.

Burge, T.: 1988a, Cartesian Error and the Objectivity of Perception. In R.H. Grimm and D.D. Merrill (eds.), *Contents of Thought*, Tucson, AZ.: University of Arizona Press, pp. 62-76. Also in P. Pettit and J. McDowell (eds.), *Subject, Thought, and Context*, Oxford: Oxford University Press, 1986.

Burge, T.: 1988b, Authoritative Self-Knowledge and Perceptual Individualism. In R.H. Grimm and D.D. Merrill (eds.), *Contents of Thought*, Tucson, AZ.: University of Arizona Press, pp. 86-98.

Crane, T.: 1992a, Introduction. In T. Crane (ed.), *The Contents of Experience: Essays on Perception*, Cambridge: Cambridge University Press, pp. 1-17.

Crane, T.: 1992b, The Nonconceptual Content of Experience. In T. Crane (ed.), *The Contents of Experience: Essays on Perception*, Cambridge: Cambridge University Press, pp. 136-57.

Davies, M.: 1889, Tacit Knowledge and Subdoxastic States. In A. George (ed.), *Reflections on Chomsky*, Oxford: Basil Blackwell, pp. 131-52.

Davies, M.: 1991, Individualism and Perceptual Content. *Mind* vol. 100, pp. 461-84.

Davies, M.: 1992, Perceptual Content and Local Supervenience. *Proceedings of the Aristotelian Society* vol. 92, pp. 21-45.

Davies, M.: 1993, Aims and Claims of Externalist Arguments. In E. Villanueva (ed.) *Philosophical Issues Volume 4: Naturalism and Normativity*, Atascadero, CA.: Ridgeview Publishing Company, pp. 227-49.

Davies, M. and Humphreys, G.W.: 1993, Introduction. In M. Davies and G.W. Humphreys (eds.), *Consciousness: Psychological and Philosophical Essays*, Oxford: Blackwell Publishers, pp. 1-39.

Dennett, D.: 1988, Quining Qualia. In A.J. Marcel and E. Bisiach (eds.), *Consciousness in Contemporary Science*, Oxford: Oxford University Press. Reprinted in W.G. Lycan (ed.), *Mind and Cognition: A Reader*, Oxford: Basil Blackwell, 1990, pp. 519-47.

Evans, G.: 1982, *The Varieties of Reference*. Oxford: Oxford University Press.

Fricker, E.: 1991, Content, Cause and Function (Critical Notice of McGinn, *Mental Content*), *Philosophical Books* vol. 32, pp. 136-44.

Fodor, J.: 1986, Individualism and Supervenience. *Proceedings of the Aristotelian Society* Supplementary Volume 60, pp. 235-62.

Fodor, J.: 1987, *Psychosemantics*. Cambridge MA.: MIT Press.

McFetridge, I.G.: 1985, Supervenience, Realism, Necessity. *Philosophical Quarterly* vol. 35, pp. 246-58. Reprinted in *Logical Necessity and Other Essays*, London: Aristotelian Society, 1990, pp. 75-90.

McGinn, C.: 1982, The Structure of Content. In A. Woodfield (ed.), *Thought and Object*, Oxford: Oxford University Press, pp. 207-259.

McGinn, C.: 1989, *Mental Content*. Oxford: Basil Blackwell.

Matthews, R.J.: 1988, Comments (on Burge 1988a). In R.H. Grimm and D.D. Merrill (eds.), *Contents of Thought*, Tucson, AZ.: University of Arizona Press, pp. 77-86.

Peacocke, C.: 1983, *Sense and Content*. Oxford: Oxford University Press.

Peacocke, C.: 1989, What Are Concepts? In P.A. French, T.E. Uehling and H.K. Wettstein (eds.), *Midwest Studies in Philosophy Volume 14: Contemporary Perspectives in the Philosophy of Language II*, Notre Dame: University of Notre Dame Press, pp. 1-28.

Peacocke, C.: 1992, *A Study of Concepts*. Cambridge, MA.: MIT Press.

Putnam, H.: 1975, The Meaning of "Meaning". In *Philosophical Papers Volume 2: Mind, Language and Reality*, Cambridge, Cambridge University Press, pp. 215-71.

Segal, G.: 1989, Seeing What Is Not There. *Philosophical Review* vol. 98, pp. 189-214.

Segal, G.: 1991, Defence of a Reasonable Individualism. *Mind* vol. 100, pp. 485-94.

Tye, M.: 1991, *The Imagery Debate*. Cambridge, MA.: MIT Press.

Tye, M.: 1992, Visual Qualia and Visual Content. In T. Crane (ed.), *The Contents of Experience: Essays on Perception*, Cambridge: Cambridge University Press, pp. 158-76.

Corpus Christi College, University of Oxford
England

WILLIAM RAMSEY

CONCEPTUAL ANALYSIS AND THE CONNECTIONIST ACCOUNT
OF CONCEPTS

I. INTRODUCTION

Recent work in connectionist modeling has had a major impact on contemporary philosophy of psychology - it has forced many philosophers to seriously rethink a number of important topics such as the structure of mental representation, the nature of information storage and the character of learning. In this essay, however, I want to link connectionism to philosophy in a different way. Rather than discuss the importance of connectionism for some philosophical *topic*, I want to focus on the implications of this research for the *way* philosophy often gets done. More specifically, I want to take a look at what many connectionists have to say about the way we represent concepts and discuss some consequences of their views for the popular philosophical enterprise of conceptual analysis - i.e., the search for precise definitions specifying necessary and sufficient conditions for abstract notions. My aim will be to convince you that if much of what the connectionists are saying is true, then analytic philosophers need to seriously rethink this popular strategy for understanding the defining many abstract concepts.

 To do this, I plan to proceed as follows. In the next section, I'll give a brief characterization of the sort of philosophical enterprise I'm calling conceptual analysis and try to make explicit some of the presuppositions that it rests upon. In Section III, I'll provide a sketch of some the more popular accounts of concept representation that are currently being endorsed by connectionist researchers. In Section IV, I'll examine the implications of the connectionist theories discussed in Section III for the philosophical enterprise discussed in Section II. Here, I'll develop and defend my claim that the assumptions which drive conceptual analysis clash with popular connectionist theories of concept representation. If what the connectionists are saying is true, then the search for crisp definitions of various abstract notions by probing our intuitions is a seriously misguided endeavor. Finally, in Section V I'll offer a brief conclusion and discuss possible ways the enterprise of conceptual analysis might be brought in line with the

35

A. Clark et al. (eds.), Philosophy and Cognitive Science, 35–57.
© 1996 *Kluwer Academic Publishers.*

connectionist research.

Before we get started, however, a couple of preliminary comments are in order. The first is that the thesis I'll defend is a conditional one concerning what follows *if* certain connectionist accounts of concept representation are correct. Since my claim is about the *relation* between this branch of empirical research and philosophical analysis, I'll make no real effort to defend the connectionist accounts themselves. Secondly, although no one as far as I know has tried to run the sort of detailed argument I'm going to present, it's worth noting that many of the themes I'll be discussing have been floating around philosophical circles for some time. Perhaps the best known expression of these sentiments is Wittgenstein's discussion of family resemblance concepts in the *Investigations*, though similar ideas can be found of the writings of other philosophers, including Hilary Putnam (1962), Peter Atchinstein (1968), Harold Brown (1988), Terence Horgan (1990) Goschke and Koppelberg (1991) and, in particular, Stephen Stich (1990, forthcoming), whose recent work help inspire this paper. The following can be read as an attempt to ground these ideas in the empirical base of an actual cognitive models and give them a full-blooded argument.

II. PHILOSOPHICAL INVESTIGATIONS OF CONCEPTS

Since antiquity (starting, perhaps, with the writings of Plato), philosophers have pursued an enterprise that is commonly known of as "conceptual analysis". For those engaged in this endeavor, the general goal has been to discover the correct analysis or definitions for a number of different abstract concepts. The range of notions investigated by this method has been quite diverse, including such concepts as *knowledge, causation, rationality, belief, person, justification* and *morality*. As one author notes, "What philosophers throughout their history have sought are those characteristics of what they were examining, whether it be *knowledge, truth, necessity, mind, recklessness, value*, or *time*, in virtue of which it is what it is; those characteristics which are necessary to it and give its essence" (White, 1975, p. 103). Among the more optimistic philosophers doing conceptual analysis, one hope has been that it would yield for philosophical notions the same sort of precise definitions that mathematicians have been able to provide for mathematical concepts. Even philosophers who don't regard the analysis of some concept as their *primary* goal, nevertheless, either explicitly or implicitly, often undertake some form of conceptual analysis as a step toward achieving some further end. Thus, conceptual analysis has been a significant part of Western philosophy and is thought by many to be the only reliable method of philosophical inquiry into the nature of a number of abstract notions. But how does this enterprise get carried out and, perhaps more importantly, what are its underlying assumptions about the way we represent concepts?

II. A. *Two Criteria for Definitions*

Answering the first question - i.e., how does conceptual analysis get done? - is, at first glance, relatively easy: philosophers propose and reject definitions for a given abstract concept by thinking hard about intuitive instances of the concept and trying to determine what their essential properties might be. However, this characterization is really too vague to tell is us anything useful. Perhaps a better way to gain insight into conceptual analysis is to consider what is normally expected of the definitions put forth. By looking at the criteria philosophers use for definitions, we can get a firmer grasp on what philosophers are up to and perhaps uncover some of the presuppositions lurking behind this enterprise.

Naturally, there are a number of different criteria commonly invoked by philosophers searching for definitions. Here, I'll focus upon only two that, although rarely mentioned, nevertheless have a major influence on the way conceptual analysis often gets done. The first of these requirements is that the definitions be relatively straightforward and simple. Indeed, a popular syntactic form assumed for definitions is that of a small set of properties regarded as individually necessary and jointly sufficient for the concept in question. Hence, philosophical definitions often take a syntactic form of "if and and only if", followed by a short *conjunction* of properties. For example, a venerable tradition in philosophy holds that X is knowledge if and only if X is justified, true belief or X is acting freely if and only if X is doing what he or she wants. As with explanatory theories in science, a popular underlying assumption of conceptual analysis is that overly complex and unwieldy definitions are defective, or ad-hocish, even when no better definition is immediately available. If an analysis yields a definition that is highly disjunctive, heavily qualified or involves a number of conditions, a common sentiment is that the philosopher hasn't gotten it right yet. Accordingly, different analyses are typically regarded as *competitors*, and, for the most part,

few people take seriously the idea that the correct analysis might be one
involving a disjunctive combination of these alternate definitions. For many
philosophers, a proposed definition should be short and simple.

A second criterion definitions are commonly expected to meet in
philosophical circles is a concern not about their form, but about the range of
cases the definition is supposed to capture. If a definition is to count as a *real*
definition, then it is generally assumed that it cannot admit of any intuitive
counterexamples. Hence, in philosophical circles it is generally assumed that
the standard way to gun down a proposed analysis is to find either a non-
instance of the concept that possesses the definitional properties in question -
thereby showing that the defining properties are insufficient to capture the
concept - or an instance of the concept that doesn't possess the definitional
properties - thereby showing the defining properties aren't necessary. If
counter-examples of this sort can be found, then the proposed definition is
typically regarded as inadequate. This sentiment is nicely expressed in the
following passage from a popular text on philosophical method:

> [W]e shall tentatively consider a definition satisfactory if,
> after careful reflection, we can think of no possible
> examples in which either the defined word truly applies to
> something but the defining words do not, or the defining
> words truly apply to something but the defined word does
> not. When we can think of such an example, then we have
> found a counterexample to the alleged definition showing
> that we do not have an accurate reportive definition. If we
> can find no counterexample to a definition, then we may
> regard it as innocent until a counterexample is found to
> prove otherwise (Cornman, et al. 1982, p. 18).

Hence, definitions sought by philosophers engaged in conceptual analysis
typically must pass at least two tests: they must be relatively simple -
generally a conjunction of individually necessary and jointly sufficient
properties, and it must not admit of any intuitive counterexamples. With this
in mind, we can now turn to the question of psychological presuppositions.

II. B: *Psychological Presuppositions of Conceptual Analysis*

At first blush, it might seem a little odd to suppose that conceptual analysis
involves *any* presuppositions about the way our minds work. After all, if

people are interested in defining notions like *justice* or *causation*, then it's justice or causation that they are concerned with - not human psychology. Nonetheless, when we look more closely at the criteria for definitions I've just sketched, we can indeed find lurking in the background certain assumptions about human cognition. Perhaps the easiest way to see this is to consider the significant role intuitive categorization judgments play in this type of philosophy. Notice, for example, that for either type of counterexample to actually count as a counterexample, there are going to have to be fairly strong and widely shared intuitions that some particular thing or event either is or is not an instance of the concept in question. In other words, the process of appraising definitions requires comparing and contrasting the definitional set of properties with *intuitively* judged instances and non-instances of the target concept. Without these intuitive categorization judgments, conceptual analysis as a practice could never get off the ground.

Because of this important role of intuitive judgments, conceptual analysis can't avoid being committed to certain assumptions about the nature of our cognitive system. One such assumption is that there is considerable overlap in the sorts of intuitive categorization judgments that different people make. Without this consensus, an intuitive counterexample for one individual would fail to be an intuitive counterexample for another individual, and no single definition could be agreed upon. Moreover, given that definitions are expected to express simple conjunctions of essential properties and allow no intuitive counterexamples, there also appears to be the fairly strong presumption that our intuitive categorization judgments will coincide perfectly with the presence or absence of a small but specific set of properties. In other words, lurking in the background of this enterprise is the assumption that our intuitions will nicely converge upon a set whose members are all and only those things which possess some particular collection of features. Given that philosophers expect to find tidy conjunctive definitions, and given that they employ *intuitions* as their guide in this search, the presupposition seems to be that our intuitive categorization judgments will correspond precisely with simple clusters of properties.

A reasonable question to ask at this stage is where do these assumptions come from? After all, *prima facie* one would suppose that it is open question whether or not our intuitive judgments can guide us to a relatively simple set of defining characteristics for abstract concepts. What exactly warrants these presuppositions?

One possible answer to this question rests with the idea that philosophers are employing an underlying folk theory of the mind or, perhaps more specifically, a folk theory about the way we represent concepts. Since Plato, a recurring theme in philosophy has been that we possess "tacit" knowledge of some domain that may not be directly accessible to our consciousness, but nonetheless is manifested through intuitive judgments and can be accessed by probing these intuitions. Similarly, outside of philosophy, Chomsky (1965, 1972) and other linguists have proposed that speakers possess tacit knowledge of their native language that takes the form of an internally represented, "psychologically real" grammar. For both Plato and Chomsky, it is recognized that this tacit knowledge is not immediately accessible to consciousness. But since this underlying competence is thought to drive our intuitive judgments, a common assumption is that it can be "uncovered" or made explicit by probing and exploiting these intuitions. For example, Chomskians assume that we can ascertain a set of syntactic rules for English by looking closely at the intuitive linguistic judgments of competent English speakers because, according to them, these judgments are generated by an actual, cognitively represented grammar of English. On this view, intuitive judgments serve as data against which we can test hypotheses about the nature of the underlying structures that produce them.

Along similar lines, it seems the easiest way to account for the assumptions that underlie conceptual analysis is by supposing that lurking in the background is a similar - though perhaps itself tacit - theory about the way we represent concepts. In other words, categorization intuitions are assumed to lead us to tidy sets of necessary and sufficient properties because, it is further assumed, these intuitions are generated by underlying representations *of* necessary and sufficient properties. On this view, it is assumed that we have tacit knowledge of the "essence" of abstract concepts, that the essence is a small set of necessary and sufficient conditions, and that we can uncover this knowledge by appealing to our intuitive categorization judgments. I'll call this theory of concept representation the "classical" view. Although relatively few philosophers engaged in conceptual analysis explicitly endorse the classical view, it strikes me as the most plausible and charitable way to make sense of this enterprise. Appealing to our intuitive categorization judgments to support or undermine a given analysis of a concept is much the same as appealing to our intuitive grammaticality judgments to support or undermine a given account of the grammar for a language. If linguists can do it, why not philosophers?

Yet, although it may be true that this view of concept representation is the most natural way to explain the assumptions of conceptual analysis, it's important to keep in mind that conceptual analysis doesn't *require* this view of concept representation. It only requires the much weaker assumption that the representation scheme for concepts - whatever form it might take - yields intuitive judgments that correspond with small sets of singly necessary and jointly sufficient conditions. While the classical view of concept representation may be the easiest way to support this weaker hypothesis, one could try to defend it by some other means. Indeed, I suppose one could avoid defending it altogether and simply take it as an article of faith that our intuitive judgments will behave in this way. Since every philosophical endeavor must start with some assumptions, this would not be a completely unreasonable strategy. It would not be unreasonable, that is, as long as there is no compelling motivation for doubting these assumptions.

The central claim of this paper is that if we take seriously the connectionist model of concept representation, then there *is* motivation for doubting both the classical theory of representation and assumptions that our intuitive judgments will yield simple and fully intuitive definitions for abstract concepts. If the sorts of stories that connectionists are now telling about concept representations and categorization judgments turn out to be correct, then the philosopher's two criteria for definitions - that they are straightforwardly simple and admit of no counterexamples - cannot be jointly satisfied. To get a better sense of where the difficulties lie, we must now turn to those connectionist accounts.

III. CONNECTIONIST ACCOUNTS OF CONCEPTS

It is now widely acknowledged that connectionist models of various cognitive phenomena depart significantly from more traditional computational theories. The connectionist account of concept representation is no exception. Actually, there is no *one* connectionist account of anything; as with any thriving research program, connectionists offer a broad range of diverse models and hypotheses. Nevertheless, it is possible to focus upon specific kinds of models if one is careful in articulating relevant distinguishing criteria. Unfortunately, many popular demarcations in connectionist literature - such as the one between symbolic and subsymbolic or localist and distributed models - fail to capture many important differences in a helpful, unambiguous way. Here, I propose to use a classification that demarcates models in terms of *what* is being represented by *which* components of the network.[1] More specifically, the sorts of models I want to look at are those in which concepts are represented by the activation patterns produced by the models' internal units. For models of this sort, there are two primary subdivisions - networks in which the individual units have an explicit semantic evaluation, and those in which they don't. I shall (rather unimaginatively) refer to the former as 'Type 1' models, and the latter as 'Type 2'.

III. A. *Type 1 Representation of Concepts*

In Type 1 connectionist models, the representation of a concept emerges out of mutual activation of individual units which by themselves encode properties generally thought to be, in some sense, component parts of the concept in question. Hence, the representation is distributed over several constituent units, each of which individually encodes some sub-element or "microfeature" of the relevant concept. The semantic interpretation of the constituent units may be just stipulated by the model builder or "acquired" by the network as a result of learning from some training set.[2] Either way, the representation of a concept in Type 1 models arises out of the simultaneous participation of several prototypical feature units. More important for our purposes, it is typically the case that in Type 1 systems, none of the individual units used to encode a feature for a given concept is necessary for the concept to become activated. Insofar as there are an adequate number of alternative feature units excited, an overall pattern corresponding to a certain concept will be produced even though a particular unit (or small number of units) may remain inert.

Perhaps the best way to get a sense of what Type 1 networks are like is by considering an actual model that embodies this structure. One nice instance of such a model is the "room-schema" network provided by Rumelhart, Smolensky, McClelland and Hinton (1986). In this model of content addressable memory, five different room concepts are represented through the mutual excitation or inhibition of 40 feature units. The connections between individual units are given either a positive or negative weight, depending upon the frequency with which objects and properties represented by their corresponding feature units are likely to co-occur, as indicated in Figure 1. For instance, the feature unit for *toaster* is positively linked with the feature unit for *refrigerator*, but negatively connected to the unit for *toilet*. Similarly, the feature unit corresponding to *fireplace* is linked

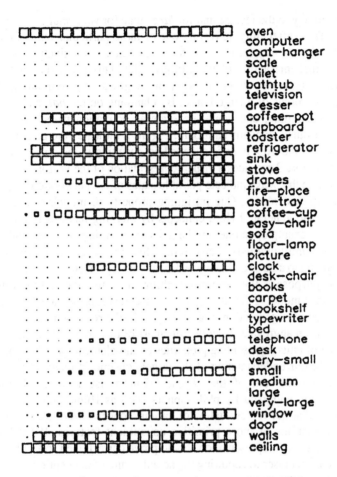

oven
computer
coat—hanger
scale
toilet
bathtub
television
dresser
coffee—pot
cupboard
toaster
refrigerator
sink
stove
drapes
fire—place
ash—tray
coffee—cup
easy—chair
sofa
floor—lamp
picture
clock
desk—chair
books
carpet
bookshelf
typewriter
bed
telephone
desk
very—small
small
medium
large
very—large
window
door
walls
ceiling

Figure 1. Weight configuration for room schema model. Each feature unit is represented by a square, indicated by the name below the square. Within each square, the small black and white squares represent the connections from that unit to the other units in the network (where the receiving unit is indicated by its relative position). Weak connections are indicated by small squares, stronger links are represented by larger squares. Positive connections are white, negative connections are black. (Source: Rumelhart, D. ET AL., 1986. Reprinted with permission from MIT Press, © 1986 by The Massachusetts Institute of Technology.)

more positively to the feature unit for *large* than for *very small*.

Thus, after fully activating one or more particular feature unit, the network subsequently settles into an overall pattern of activation as a result of the mutual interaction of the various units. This gives rise to different large-scale patterns which to varying degrees approximate different prototypical rooms of a large house. By clamping on the *bathtub* and *small* nodes, for example, one would generate an activation pattern that would represent a fairly standard bathroom. Figure 2 indicates the activation of a pattern where the *oven* node is clamped on, resulting in the representation of a kitchen. Figure 3 provides a visual representation of the vector space of this model, where the peaks indicate possible states of the network corresponding to three types of rooms. One intriguing property of models of this sort is their ability to generate hybrid concepts by merging features in an unusual manner. For example, on one run the units representing *bed*, *sofa* and *ceiling* were all simultaneously activated. This produced a representation corresponding to a large and fancy bedroom with a fireplace.

For our purposes, the important thing to note about this kind of connectionist model is that the style of conceptual representation is one that depends upon the generation of a *prototype*. In the room schema model, the prototype is an aggregate structure which arises out of the mutual activity of number of units representing properties or sub-elements which are thought to comprise our understanding of the target concept. In many ways, such a model is reminiscent of the classical Empiricist account of concept representation endorsed by philosophers such as Hume and Locke, who maintained that our representation of various concepts were constructed out of representations of more primitive constituent features. If the Type 1 account of concept representation should prove correct, then folks like Hume will have been closer to capturing the actual nature of concept representation than many have thought.

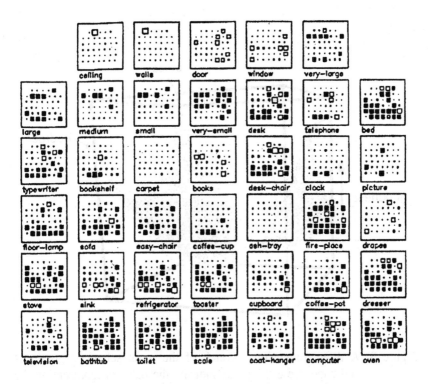

Figure 2. Activation of room schema network with *ceiling* and *oven* nodes
 fully activated (degree of activation of individual units is indicated
 by number and size of corresponding squares). Here the emergent
 pattern would represent a prototipical kitchen. (Source: Rumelhart,
 D. ET AL., 1986).

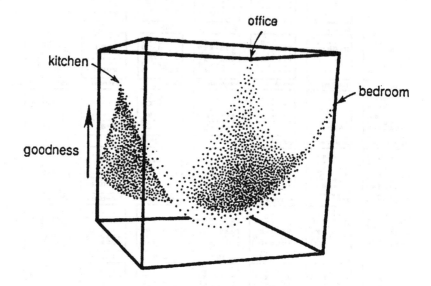

Figure 3. A three dimensional sketch of the room schema's goodness
        landscape. Here the peaks indicate the patterns representing the
        prototypical office, kitchen and bedroom. (Source: Rumelhart, D.
        ET AL., 1986).

III. B. *Type 2 Representation of Concepts*

As with Type 1 models, Type 2 models represent concepts in a distributed
fashion through the mutual activation of several individual units. The key
difference between the two is that while the constituent units of Type 1
networks admit of a semantic evaluation (i.e., they each stand for some thing
or property), the individual encoding units of Type 2 admit of no
straightforward semantic interpretation. Individually, the internal units of
Type 2 networks represent nothing. As McClelland, Rumelhart and Hinton
put it, "the internal representations are distributed and it is the *pattern* of
activity over the hidden units, not the meaning of any particular unit that is
important[3] . . . The units in these collections . . . may have no particular

meaning as individuals.".[4] It is often the case that networks of this sort are trained up by employing a learning algorithm such as back-propogation. After training, the representational capacities of these sorts of models are often uncovered by vector analysis of the models internal (or "hidden") units. Such analyses often reveal a clustering of internal activation states, suggesting that although no individual unit is used to encode a particular piece of information, different activation patterns by the same units are employed to represent different concepts.

One network that is commonly used to illustrate this type of model is Gorman and Sejnowski's mine detector. By using the popular learning algorithm, back-propogation, this three-layered feed-forward[5] network eventually acquires the capacity to discriminate the undersea sonar echoes of metal cylinders (imitation mines) from the echoes of submerged rocks. This model is represented in Figure 4a. Post-training vector analysis reveals that rock echoes tend to cluster around one region of vector space, while mine echoes tend to cluster around a different region, as indicated in Figure 4b. This has led authors such as Paul Churchland to conclude that the network has acquired a conceptual framework for recognizing and classifying the different types of echoes. According to Churchland, this classification scheme is based upon the use of prototypes: "The training process has generated a *similarity gradient* that culminates in two 'hot spots' - two regions that represent the range of hidden-unit vector codings for a *prototypical* mine and a *prototypical* rock" (1989, p. 204).

Hence, we can regard these regions in vector space as conceptual prototypes and activation patterns that fall within (or close to) such areas as representations of the relevant concept. Although Type 2 representations don't invoke sub-features, they do share with Type 1 representations a style of representing concepts that relies heavily on the use of prototypes.

As should be clear even from this brief sketch of connectionist theories of concept representation, both types pf connectionist models depart significantly from the classical account of concept representation discussed in Section II. In both Type 1 and Type 2 models, similarity to a prototype is what drives instantiation judgements - not some set of essential, defining properties. Consequently, the connectionist accounts discussed here are quite similar to many of the prototype theories of concept representation suggested in recent years by cognitive psychologists such as Eleanor Rosch (1973, 1978), Edward Smith and Douglas Medin (1981), and many regard connectionism as providing the implementational architecture for such

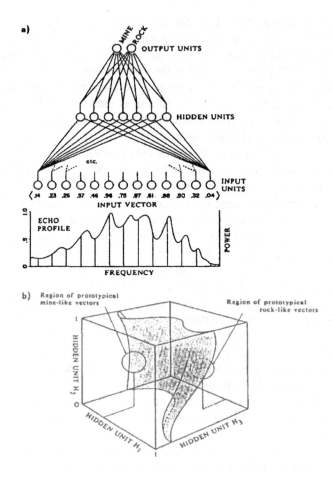

Figure 4. a) the mine/rock sonar echo discriminating network. Along with a 34 input and 2 output units, the network also has a hidden layer of 14 intermediary units. Also shown is the sonar echo profile used as the system's input. b) The learned partition on the hidden unit vector-space. For simplification, the axes are shown for only three of the seven hidden unit activation levels. (Source: Churchland, P. 1989. Reprinted with permission from MIT Press, © 1989 by The Massachusetts Institute of Technology.)

theories.[6] In a similar vein, some have invoked the Wittgensteinian language of family resemblance relations to characterize the connectionist theory of concepts. For instance, Paul Smolensky notes that, "if you want to talk about the connectionist representation . . . in this distributed scheme, you have to talk about a *family of distributed activity patterns.* What knits together all these particular representations . . . is nothing other than a type family resemblance" (1988). If these sorts of connectionist theories of concept representation are correct, then our common-sense categorization judgements about what does or does not count as an intuitive instance of some concept will be determined by such a prototype or family resemblance scheme. What we must now focus upon is the bearing all of this would have on the philosophical practice of conceptual analysis. It is to this topic that I now turn.

IV. CONNECTIONISM AND CONCEPTUAL ANALYSIS

With the groundwork completed for establishing the tension between connectionism and conceptual analysis in the last two sections, we are now in a position to see just how the argument goes. In Section II, I suggested that a possible underlying motivation for conceptual analysis is the classical view of concept representation - that we represent abstract concepts by representing a small set of individually necessary and jointly sufficient properties. Since the sort of connectionist theories of concept representation just sketched directly contradict classical accounts, it might be thought that this alone is enough to show the incompatibility of connectionist prototype theories and conceptual analysis. But as I noted in Section II, conceptual analysis doesn't actually *require* the classical theory be correct. Hence, if we are to develop a case against conceptual analysis on the basis of connectionist theories, then we must focus upon the relevance of those theories to the presuppositions upon which conceptual analysis depends.

As we saw in Section II, what the philosophical practice of conceptual analysis *does* need is the assumption that our intuitive judgments about categorization and class membership can yield a definition for abstract concepts that 1) takes the form of a small conjunctive set of essential properties and 2) coheres with all relevant intuitions and allows no counterexamples. But if the sort of theories of concept representation discussed in the last section are correct, then it is highly doubtful that both of these criteria for definitions could ever be jointly satisfied. The reason is that, *ex hypothesis*, our categorization judgments are subserved by a taxonomic scheme that will generate categorization intuitions that are too variegated and diverse to be captured by simple and non-disjunctive definitions. If being an intuitive instance of X is simply a matter of having a cluster of properties that is sufficiently similar to some prototype representation, and if there are a number of different ways this can be done, employing different combinations of properties, then any crisp definition comprised of some subset of these properties and treating them as necessary and sufficient is never going to pass the test of intuition. It will *always* admit of intuitive counterexamples because the range of diversity sanctioned by our conceptual representation scheme will be much greater than that allowed by any tidy, straightforward definition. Hence, the search for a simple, non-disjunctive definition of a given philosophical concept that accords with all of our intuitions and admits of no counterexamples is a hopeless enterprise - there simply is no such thing.

To see this more clearly, consider first Rumelhart, *et. al.*'s Type 1 Room Schema model. The representation of the concept *office* is a pattern of activation that involves the activation of a number of prototypical microfeature nodes, such as those representing *desk* and *coffee-cup*. Notice, however, that in this scheme it is possible to activate the concept *office* without activating all of the constituent microfeatures. Hence, the concept's representation admits of degrees, depending on the strength and number of microfeature units activated. More important for our purposes, different combinations of microfeatures can serve to activate the concept. For example, one representation of the concept *office* may result from the *computer, ash-tray, books, desk*, and *telephone* feature units all being activated. Another may involve *ash-tray, books* and *desk* but replace *computer* and *telephone* with *typewriter* and *coffee-cup*. Thus, none of the individual microfeatures are necessary for the representation of an office (except, perhaps, those for non-discriminating properties of all rooms such as *walls* or *ceiling*).

Now consider a philosopher's attempt to discover a definition for such a concept in accordance with the expectations of conceptual analysis. The goal would be to find a definition of individually necessary and jointly suffient properties that admits of no intuitive counter-examples. But since the concept has a disjunctive representation, any one of a number of different sets of properties will suffice to generate a judgment of instantiation. Hence, we would always be able to discover or invent intuitive instances of the concept that don't possess all of the alleged necessary properties given in any simple conjunctive definition. On the connectionist view of concept representation, intuitive counterexamples will always be forthcoming for any definition specifying essential properties.

A similar point can be made regarding the Type 2 style of concept representation, though not so directly. Recall that on this model, concepts do not have a composite structure made up of constituent microfeatures. Thus, classification judgements of concept membership are not guided by even a disjunctive set of constituent features. Instead, for something to be judged as an instance of a given concept, it must generate an activation pattern which is sufficiently close to some prototypical pattern (or perhaps some range of prototypical patterns) that corresponds with the concept's "hot spot" in vector space. Given this, it is far from clear just how, exactly, the possession of certain features possessed by the different stimuli translate into activation patterns occupying regions of vector space. But there is certainly no reason to suppose that the network classifies stimuli by invoking necessary and sufficient conditions. Indeed, when one notes that a relatively diverse range of positions in vector space will suffice for a positive categorization judgement, something very different appears to be at work. For instance, two stimuli generating hidden unit activation patterns opposite one another relative to the prototype zone would still both be judged as intuitive instances of the target concept as long as they are both close enough to that zone. In other words, two instances of a given concept can be markedly different from one another, as long as their corresponding hidden unit activation patterns fall within a certain region suitably close to the prototype. Since such a region is demarcated by a similarity gradient and not necessary and sufficient conditions, concept membership would appear to be guided more along the lines of a family resemblance model. As Paul Churchland describes matters,

> a single prototypical point or activation vector across the hidden units represents a wide range of quite different possible sensory activation patterns at the input layer: it represents the extended family of relevant (but individually perhaps nonnecessary) features that collectively unite the relevant class of stimuli into a single kind. Any member of that diverse class of stimuli will activate the entire prototype vector at the hidden units. Also, any input-layer stimulus that is relevantly *similar* to the members of that class, in part or in whole, will activate a vector at the hidden units that is fairly close, in state space, to the prototype vector (1989, p 206).

It appears, then, that a cognitive system whose categorization judgements are guided by the Type 2 mode of concept representation would likely find an analysis of a concept that appeals to essential properties inevitably unintuitive. As with Type 1 models, a characterization of concept membership in terms of necessary properties would admit of intuitive counterexamples because the stimuli producing a positive instantiation judgements are linked not by any essential features.

Thus far, my discussion has focused upon the inadequacy of definitions specifying necessary conditions. But what about definitions specifying only sufficient conditions? *Prima facie*, one would think that if either version of the connectionist story were correct, it should at least be possible to specify sufficient conditions for a given concept without having to worry about counterexamples. While the prototype and family resemblance accounts may not involve any *necessary* properties, it certainly does seem to allow that various clusters might always prove *sufficient* for instantiation judgments of the concept. Nevertheless, there are aspects of the connectionist prototype accounts which suggest that counterexamples to this sort of analysis are relatively cheap too. One such aspect that has been discussed by a number of authors is the context dependency of categorization judgements.[7] If the salience of features for categorization judgments is context sensitive, then it should be possible to find or construct examples embedded in certain circumstances where the allegedly definitional features are all present but, because of the context, are not sufficiently close to the prototype for an instantiation judgement.

It is important to see that the incompatibility of connectionism and conceptual analysis is not due to the intuitive counterexamples not being actual intuitive counterexamples to the proposed definitions. Rather, the point is that from the outlook of connectionist views of concept representation, counterexamples to classical definitions are overly cheap and abundant. While the counterinstances do indeed reflect the inadequacy of the proposed definition, it is not for the reasons traditionally assumed by practitioners of conceptual analysis. The definition is faulty not because there is some other simple set of necessary and sufficient conditions which can do a much better job of defining the concept in question. Rather, from the connectionist perspective the definition fails because *there is no simple, tidy collection of properties that is possessed by all and only intuitive instances of the concept; any definition expressed in this way is doomed from the outset.* If the connectionist account is correct, it just isn't possible, to come up with a definition that is both a simple conjunction of "essential" properties and, at the same time, captures all of our intuitions.[8]

V. CONCLUDING REMARKS

Before finishing, it's worth reflecting for a moment on how analytic philosophy itself has involved hints that something like the connectionist models discussed here might be right. The failure of conceptual analysis to produce an uncontroversial, completely satisfactory definition of the vast majority of abstract concepts should suggests that something is amiss with the classical picture. Moreover, even within the philosophical literature there have been cases suggesting difficulties with intuitions of the sort predicted by the connectionist view.[9] As noted earlier, it is far from clear what warranted the assumption that our intuitive categorization judgments should converge in a way that would produce a clean set of necessary and sufficient conditions. Since this is an empirical question concerning matters of cognition and conceptual representation, it certainly isn't something that lends itself to a priori proof.

But now what should we make of conceptual analysis if one of the connectionist models should prove to be correct? As I've tried to show, if either of the connectionist stories are correct, then the two primary expectations philosophers typically place on analyses of concepts - simplicity and avoiding all counter-examples - cannot both be satisfied. Hence, at the very least it looks like one of these criteria will need to be abandoned, or at least significantly relaxed. On the one hand, philosophers who demand analyses in the form of conjoined 'essential' properties will need to abandon the assumption that such an analysis will accommodate all our intuitions and be willing to accept the existence of intuitive counter-examples. On the other hand, if philosophers are going to insist on air-tight analyses that admit of no intuitive counter-examples, then they must abandon the hope for tidy definitions and develop a tolerance for highly disjunctive and heavily qualified definitions. Naturally, whichever strategy philosophers adopt will depend a good deal on the sort of work we want our analyses to do for us. Yet even here, in deciding which of the two criteria we ought to relax, connectionism may have an important role to play. For example, Barsolou (1987) has argued that there is considerably more variance in categorization judgements among different individuals, and that in any given individual such judgements tend to change over time. Barsolou believes that one plausible explanation for this is that our concept scheme is embodied by connectionist networks similar to those discussed here, but that lack a static structure. If such a model should prove correct, then any fixed definition - even one that is highly disjunctive - may fail to capture all of the intuitions all of the time since these intuitions would not be universal or static. Consequently, abandoning simplicity in order to accommodate all intuitions may be a misguided strategy, and a more promising approach might be the one of preserving simplicity and recognizing that we can't capture all intuitive judgments. It's worth noting that this latter tactic is not entirely foreign to philosophy. For instance, in certain respects it is quite similar to Carnap's (1950) eloquent discussion of concept *explication*, where *revision* of the our intuitive categorization judgments is explicitly endorsed for the sake of simplicity. Similarly, in response to the various intuitive counter-examples to utilitarianism, Smart (1965, 1990) suggests that it is our intuitions, and not the utilitarian account of morality, that should be questioned or ignored. If the connectionist story of concept representation is correct, then the strategies suggested by these philosophers - in particular, decreasing the importance of intuitive counterexamples - merit more consideration and application than they have received.[10]

NOTES

[1] For a more detailed elaboration of this classification system, see Ramsey (1992b).

[2] There are, of course, a number of philosophical worries associated with claims about parts of artifical systems "acquiring" semantic content without some outside stipulation. Fortunately, these matters need not occupy us here. For a brief attempt to elaborate on this topic, see Ramsey (1992b).

[3] In Rumelhart and McClelland (1986), p. 344-346.

[4] Rumelhart and McClelland (1986), p. 33.

[5] A feed-forward network is one where the activation spreads in only one direction -- from input to output.

[6] See, for example, Churchland (1989), Goschke & Koppelberg (1991) and Smolensky (1988).

[7] See, for example, Smolensky (1988).

[8] One objection that has been raised to this sort of attack on conceptual analysis is that it confuses the metaphysical issue of a concept's actual extension with the epistemological issue of how we represent categories (Rey, 1983, 1985). The problem with this move, as I see it, is that it is actually the advocate of conceptual analysis who supposes that we can uncover the true essence of philosophical notions by simply probing our intuitive judgements. Hence, it is the proponent of conceptual analysis, and not me, who forges a strong link between concept definitions and the way we represent concepts. For a more

complete discussion of this objection and my response, see Ramsey (1992a).

[9]   One example is Bernard Williams' (1970) well-known discussion of personal identity. Here, a proposed set of conditions for personal identity yields positive intuitive judgments when described in one way, but, as Williams points out, basically the same set of conditions tend to elicit contrary judgments when described somewhat differently. This is, of course, just what one would expect if the categorization judgments are context-sensitive in the way suggested by certain connectionist theories.

[10]   Portions of this essay appear in Ramsey (1992b). Earlier versions were presented at Central Michigan University and the University of Memphis. A great deal of useful feedback was provided by these audiences.

REFERENCES

Achinstein, P.: 1968, *The Concepts of Science*, John Hopkins University Press, Baltimore, Maryland.

Barsolou, L.: 1987, 'The instability of graded structure: implications for the nature of concepts', in *Concepts and Conceptual Development: Ecological and Intellectual Factors in Categorization*, U. Neisser (editor), Cambridge University Press, Cambridge.

Brown, H.: 1988, *Rationality*, Routledge, London.

Carnap, R.: 1950, *Logical Foundations of Probability*, University of Chicago Press, Chicago.

Chomsky, N.: 1965, *Aspects of the Theory of Syntax*, MIT Press, Cambridge MA.

Chomsky, N.: 1972, *Language and Mind*, Harcourt, Brace and Jovanovich, New York.

Churchland, P.: 1989, *A Neurocomputational Perspective*, MIT/Bradford, Cambridge, MA.

Cornman, J. W., Lehrer, K., and Pappas, G. S.: 1982, *Philosophical Problems and Arguments: An Introduction* (3rd Edition), Macmillan, New York.

Goschke, T. and Koppelberg, D.: 1991, 'The concept of representation and the representation of concepts in connectionist models', in *Philosophy and Connectionist Theory*, W. Ramsey, S. Stich and D. Rumelhart (editors), Lawrence Erlbaum, Hillsdale, NJ, 129-161.

Horgan, T.: 1990, 'Psychologistic semantics, robust vagueness, and the philosophy of language', in *Meanings and Prototypes: Studies in Linguistic Categorization*, Routledge, London, 535-557.

Rey, G.: 1983, 'Concepts and stereotypes', *Cognition* 15, 237-262.

Rey, G.: 1985, 'Concepts and conceptions: A reply to Smith, Medin and Rips', *Cognition* 19, 297-303.

Ramsey, W.: 1992a, 'Prototypes and conceptual analysis', *Topoi* 11: 59-70.

Ramsey, W.: 1992b, 'Connectionism and the philosophy of mental representation', in *Connectionism: Theory and Practice*, S. Davis (editor), Oxford University Press, Oxford, 247-276.

Rosch, E.: 1973, 'On the internal structure of perceptual and semantic categories', in *Cognitive Development and the Acquisition of Language*, T. E. Moore (editor), Academic Press, New York.

Rosch, E.: 1978, 'Principles of categorization', in *Cognition and Categorization*, E. Rosch and B. Lloyd (editors), Lawrence Erlbaum, Hillsdale, New Jersey, 27-48.

Rumelhart, D., Smolensky, P., McClelland, J. and Hinton, G.: 1986, 'Schemata and sequential thought processes in PDP models', in *Parallel Distributed Processing*, Vol. II, MIT/Bradford, Cambridge, MA, 7-57.

Smart, J. J. C.: 1965, 'The methods of ethics and the methods of science', *Journal of Philosophy* **62**, 344-349.

Smart, J. J. C.: 1990, 'Integrity and squeamishness', in *Utilitarianism and its Critics*, J. Glover (editor), Macmillan, New York, 170-174.

Smith, E. A. and Medin, D. L.: 1981, *Categories and Concepts*, Harvard University Press, Cambridge, MA.

Smolensky, P.: 1988, 'On the proper treatment of connectionism', *The Behavioral and Brain Sciences* **11**, 1-74.

Stich, S. P.: 1990, *The Fragmentation of Reason*, Bradford/MIT Press, Cambridge, MA.

Stich, S. P.: (forthcoming), 'What is a Theory of Mental Representation?'.

White, A. R.: 1975, 'Conceptual analysis', in *The Owl of Minerva*, C. J. Botempo and S. J. Odell (editors), McGraw-Hill, New York.

Williams, B.: 1970, 'The self and the future', *Philosophical Review*, Vol. **LXXIX** (2), 161-180.

Wittgenstein, L.: 1953, *Philosophical Investigations*, Macmillan, New York.

William Ramsey
University of Notre Dame

ANDY CLARK

OF NORMS AND NEURONS *

I. INTRODUCTION/ABSTRACT

What has philosophy to say about the internal structure of intentional
systems -systems properly characterised as believing that so-and-so, hoping
that such-and-such etc. etc.? One radical view, canvassed by the present author
in Clark (1991) is "Nothing". According to that treatment, any being whose
behavioural patterns are appropriately complex can count as having beliefs
etc. regardless[1] of the nature of the inner processing involved. The present
paper suggests that such a view is inadequate insofar as it fails to take
account of a requirement which I shall label 'normative depth', viz.

> (Requirement of Normative Depth)
>
> The inner workings of an intentional system must be of a
> kind compatible with the description of that system as
> capable of making mistakes which involve the failure to
> respect those commitments in episodes of on-line
> processing.

This turns out to be an interesting requirement in that it is both weak enough
to allow for a wide variety of acceptable inner structures (certain kinds of
connectionist and/or artificial life architectures, as well as classical ones) yet
strong enough to rule out one of the great 'evil demons' of contemporary
discussion - the Giant Look-up Table.

The structure of the discussion is as follows. Section 1 introduces
two (related) philosophical constraints on intentional systems (the Generality
Constraint and the aforementioned Requirement of Normative Depth). Section
2 applies these constraints to the case of the Giant Look-up Table. This far-
fetched example is then replaced (Section 3) by the case of a connectionist
network. I show that only relatively advanced forms of connectionist model
will meet the constraints. In Section 4 I briefly consider the radical
architectures developed by Rodney Brooks (e.g. Brooks (1987)) and conclude
that even these non-representational approaches could, suitably developed,

59

A. Clark et al. (eds.), Philosophy and Cognitive Science, 59–71.

meet the requirements we have set. I conclude (Section 5) that the Normativity requirements are well placed to play a crucial role in establishing the proper intersection between philosophical and architectural concerns.

II. TWO CONSTRAINTS.

2.1 *The Generality Constraint*

Consider the belief ascription "Andy believes that pigeons are not migratory birds". This ascribes to me a structured mental content. A plausible condition of enjoying such a content is that I be master of the individual concepts involved e.g. that I be master of the concept 'pigeon', 'migration', the negation operation etc. But to grasp a concept is to grasp its intended width of application - to see it as a potential modifier of an infinite (or at any rate very large) number of other thoughts. Thus I must grasp the general concept of migration, which involves knowing that it is a term properly applicable to any animal who comes and goes with the seasons. Likewise, I must know that pigeons are the proper subject of many other predicates. In short, I must be able to deploy my cognitive competence with the concepts in a very wide range of cases, showing the proper interanimation with all the other concepts of which I am master. This is, in essence, the so-called Generality Constraint of Evans (1982).

2.2. *Normative Depth*

This is the constraint which I will focus on. It is intended to cover two central features of our mental life. One is our ability to distinguish the accidental from the constitutive, especially in the context of concept-application. The other is our ability to judge our own performance as either living up to our antecedent commitments, or as failing to live up to them. We can illustrate both points by borrowing an example from Martin Davies. Consider the concept 'grandmother'. As a master of that concept I command both a stereotypic picture of a grandmother (an old grey-haired lady who likes a small gin, perhaps), and a constitutive rule (any mother of a parent is a grandmother). It is central to my grasping the concept that I realise that the constitutive rule is in the driving seat. If Jane Fonda is the mother of a parent, then she ought to fall under the concept 'grandmother', accidental characteristics notwithstanding. This is where the issue of judging our own performance comes in. For I may, in daily commerce, easily fail to apply the concept 'grandmother' to non-stereotypic cases (e.g. fail to notice that Fonda is eligible for some welfare benefit). In such cases it must be possible for me to step back and say "I can see, given my own canonical commitments, that I ought to have judged thus and so". This ability to recognise our own guiding cognitive commitments and to judge our own judgings as correct or mistaken is, I suspect, central to the idea of moral agency. And it is plausible to suppose that part of the motivation for a concern with a proper membership of the class of believers, is a moral one, viz. that only those beings are properly assessable for their deeds and judgements.

The two constraints, thus described, seem plausible enough. At any rate, I do not propose to argue further for them here. The question I really want to address, and to which we now turn, is rather this: supposing we are drawn to some such constraints, what (if anything) do they imply about the internal organisation of genuine intentional systems?

III. UNMASKING THE PRETENDERS.

Consider the kind of internal organisation exhibited by a giant look-up table. It contains, for every possible input (located in a historical series) a canned output behaviour. How does it fare with respect to the constraints just discussed?

The constraint of concept-grasp is met (naturally) on the surface. Its verbal behaviour is indistinguishable from yours and mine. But does that behaviour rest on an interlocking and interanimated set of cognitive competencies? It seems clear that it does not. Every case is treated as a special, individual, unique case. The fact that it 'knows' what to say when asked "Do pigeons fly?" and the fact that it knows what to say when asked "Have you ever eaten pigeon?" are organisationally independent. By this I mean that there are no causally interlinked informational resources which are implicated in each transaction and which became causally unified because of the content overlap between the cases. This idea of internal causal links being created because of the semantic relations between contents is the key internal constraint on which I shall be relying. Any system in which apparent concept-grasp is attained by fully independent canned responses fails to meet such a condition. Thus the constraint of concept-grasp is immediately (see also Davies (1991)) a constraint upon innards. Notice, however, that in insisting upon such semantically determined causal interanimation, we need not insist upon the propositional character of the internal objects thus interanimated. All we require is that a variety of internal states, which have formed in response to semantic properties of inputs, turn out to be in internal causal commerce because of those semantic commonalities[2]. Basic abilities of cross-modal integration (the partially common internal causal effects of a fire engine seen and heard) satisfy the demand.

What, then, of the normativity constraint? Once again, a similar move recommends itself. The look-up table meets a behavioural criterion of normative depth. But examination of its innards should force us to think again. Consider the case of the non-stereotypic grandmother. Given an appropriate input (e.g. someone says, "Look, you've missed Jane Fonda. She's a granny too") the look-up system outputs "Oh yes. I made a mistake. Sorry." But that output is not the result of causal links formed between the states which underlie its stereotypic and constitutive knowledge of grandmothers, and formed because of the semantic commonality in their contents. Once we see this, I suggest, we see that its behaviour, though surface-indistinguishable from mine, is not genuinely normative. It is not the failure of fit of the earlier output to the being's canonical commitments which is internally responsible for the revised judgement.

The look-up table thus fails to meet an exploded version of the conceptual requirements detailed in section 1. It fails to meet the basic demands of conceptual interanimation since its innards exhibit no causal relations whatsoever between the various internal states which ground its apparently unified and general competence with e.g. pigeon-thoughts. And it fails to meet the advanced demands of normative depth since its innards lack the self-monitoring structure necessary to underwrite the claim that some revised judgement (e.g. recognition that a previous inference was mistaken) was brought about by a comparison of its earlier output to the one determined by its canonical commitments.

So far, though, we have relied on a very far-fetched example of an internal organisation. To get a sense of how our requirements constrain a more realistic architecture, I propose to indulge in a brief foray into the frontiers of cognitive science. The aim of the journey is simply to show that the constraints have bite even for more interesting organisational structures than (mere) look-up tables.

IV. NORMS AND NEURAL NETWORKS.

The most revealing swift characterisation of connectionist cognitive models is probably this; such models eschew the direct manipulation of strings of a syntactically structured semantically interpretable inner code. They instead relay heavily on the prototype-style representations formed by statistical feature extraction from a well-chosen training set. It would be wasteful to rehearse much of the detail here (see e.g. Clark (1989), Churchland (1989)). Instead, let us consider one example.

Imagine a network whose task is to mediate grandmother recognition. It is trained by giving it a series of examples (some of grannies, some not) and allowing an automatic procedure to adjust its internal characteristics (the weights on the links between idealised 'neurons' or units) towards a successful granny-spotting configuration. The examples will be specified as a set of features e.g. granny-1 might be specified as grey-haired, gin-drinking and 70, granny-2 might be blonde-haired, gin-drinking and 60 and so on. After training, the network should have a fair sense of the statistical centre of the space of example-grannies to which it has been exposed. That is to say, it will judge new cases to fall under the concept according to their relative distance from the common trend or trends of the examples. This is a very powerful procedure, and has been used, for example, to train a network to discriminate rocks from mines on the basis of sonar echoes (to a sensitivity somewhat in excess of that of a good human sonar operator (see Churchland (1989)).

We may immediately note that the basic mode of representation in such systems, though quite different to classical, Language of Thought style schemes (see e.g. Fodor (1987), Fodor and Pylyshyn (1988)) is quite able to satisfy the demand (1.1 and 1.2 above) that the inner states underlying various apparent manifestations of grasp of a single concept should not be fully internally disunified. For such states will be the various (context-sensitive) representations constituted by a variety of hidden unit activation vectors in a connectionist network. Such states exist as the partially overlapping patterns of hidden unit activation which such systems develop to deal with semantically related circumstances. The overlap in the use of internal resources (see e.g. Smolensky (1991)) is a direct result of the semantic commonalities between the cases. (By contrast, connectionist systems with a surfeit of hidden units sometimes devote entirely separate resources to each case - such networks fail to meet our demands: they are just the connectionist analogues to look-up tables.) A well balanced network will thus meet our internalist version of the generality constraint very nicely, and, without embodying anything like a language of thought.

Things are not so simple, however, when we consider our second constraint viz. that the networks exhibit genuine normative depth. To see why, recall a central drawback (made much of by e.g. Pinker and Prince (1988)) that such networks look doomed to form only example-bound knowledge. Thus if, for example, a network's knowledge of the grammatical rule 'add -ed to the stem to form the regular past-tense' is just a matter of the statistical trend common to a class of example regulars, it won't be able to deploy that rule in its full generality. For if a novel verb is very distant from the centre of the space of particular examples, it will not know to give it the regular ending. Human implicit knowledge of that rule seems, by contrast, to be fully general and abstract. Likewise, if a putative granny-description is far from the centre of the space of examples (as Jane Fonda could easily be) a standard network will fail to apply the concept. What is required in both cases turns out to be a more complex internal architecture than a single network of the standard 3-layer pattern associator kind typical of 'first wave' connectionism.

To get a sense of this increasing complexity, consider two relatively recent proposals which might help ease the situation just described. The first is a technique known as skeletonization. One problem with pattern associators is that they tend to extract all the statistical regularities from the example class and hence form no idea of the minimal features necessary to determine category membership. It will be evident that these minimal features are (in the grandmother case) nothing other than the constitutive feature itself viz. being the mother of a parent. Mozer and Smolensky (1989) describe an automatic procedure which takes a trained net (e.g. one which is already quite successful at performing the task) and computes a measure of relevance for each unit. It then destroys such units as were least necessary to yield correct outputs. Thus on a categorisation task involving both a sufficient rule and a number of partial statistical correlations, the reduced or skeletonized network was able to retain only the units whose activity was correlated with the input features which figured in the sufficient rule. The skeletonized network thus transcended the mere statistical sensitivity of its parent to utilise the minimal feature set which defined a category. The minimal feature set which defines the category grandmother, of course, is just the set with the unique member 'is a mother of a parent'. Skeletonization of rich conceptual resources thus looks like a step towards distinguishing constitutive from accidental features within a connectionist framework.

It is only a step, however, since a universally effective but accidental indicator would pass the skeletonization test (e.g. if every granny as a matter of fact was gin-drinking). What we really need is not merely a further statistical procedure (which is what skeletonization amounts to) but an integrated cognitive economy in which more theoreticised network representations are able to (on occasion) override the judgements of networks working on more low-grade or perceptual data. In short, we require a functional hierarchy of representations with tight causal relations between the levels. It is quite possible, for example, to have a network whose state space is devoted to genealogical relations. (For a convincing argument that such highly theoreticised contents are amenable to connectionist, prototype-style representation see Churchland (1989)). What is required then is that in certain circumstances such a network be capable of 'taking the causal reins' and issuing the system's final judgement "Fonda is a grandmother" even though the less theoreticised sub-networks, left to their own devices, would disagree. Such multi-level connectionist models (examined in some detail in Clark and Karmiloff-Smith (forthcoming) and more concretely in Legendre, Miyata and Smolensky (1990a) and (1990b)), make room for the distinctive normative phenomenon in which our canonical commitments outrun our on-line performance. (See also Davies (1991).) Thus in a day to day context it might be the less theoreticised nets which mediate granny-recognition. On-line errors might then be judged as such by the same overall system when it is prompted to deploy its higher level resources. (Note that in line with the organisational maxims of section 2 we must insist that the system has forged the causal pathways which allow the higher-level override because of the semantically common subject matter involved i.e. because the common object of the perceptual and more theoretical networks is, precisely, grandmothers).

This brief detour, then, is meant to have shown both that standard connectionist systems find it difficult to meet the requirement of normative depth and that meeting it involves sophistications to the organisational structure of the system. Once networks are set up so as to exhibit the required depth, we typically find an internal organisation which begins to satisfy the internal structural requirements outline earlier[3].

V. ROBOTS AND REPRESENTATIONS.

It is worth asking just how far we can depart from the familiar symbolic paradigm once we are confined to produce systems with conceptual internanimation and normative depth. To some extend, this is a merely empirical issue. There are certainly those who feel that what is distinctively connectionist will have to be lost as we approach the higher cognitive realms (see e.g. Fodor and Pylyshyn (1988)). But at least there is nothing (or so I shall now argue) intrinsic to our organisational demands which depends on the manipulation of inner syntactic items which are in any way recognisable as classical symbol structures.

To see why, let us introduce the idea (devised by Rodney Brooks - see Brooks (1987)) of an abstraction barrier. An abstraction barrier is a stage in a system's information processing where a task (e.g. planning) is performed using a symbolic code. Thus consider (Brooks' example) a mobile robot who must negotiate some small area of physical terrain. One way of achieving this is to have a cognitive module which plans a route in explicit, symbolic form and then passes commands to a lower level program which (let's say) moves the wheels. In Brooks' terms the lower level program then simply 'does the right thing' i.e. it directly responds to the action specification by controlling the requisite motors. But, says Brooks, "one could imagine ... yet another level of planning, where an explicit model of the drive circuits was used to symbolically plan how to vary the motor currents and voltages!". (Brooks (1987), p. 3) Brooks argues that by pondering such absurdity we can begin to see that the whole idea of an abstraction barrier is in effect an artifact of an overly rationalistic model of intelligence. At every point, he believes, we have a choice between inserting a symbolic planner or inserting a program which just 'does the right thing'. What this means in practice is short-circuiting the classical sense-think-act cycle (see Malcolm, Smithers and Hallam (1989)) by setting up tight sensitive feedback loops between sensing and action. Some quite impressive results in simple robotics[4] have been achieved by thus entirely eschewing the notion of internal manipulation of symbolic models.

Would such an approach fall foul of the kind of organisational demand we have been considering? Surely not. For all that we have demanded is:

(a)   that the internal states which underlie the various
             uses of a concept should be causally interrelated.

(b)   that they should have come to be so related in
               virtue of internal self re-organisation caused by
               the semantic commonality between the cases,

and

(c)   that the internal organisation should allow the
               system genuinely to judge its own performance
               (which in turn involves real causal links between
               the states underlying the original performance
               and those underlying the judgement - a condition
               not met by e.g. the look-up table).

These requirements are positively congenial to the idea of dissolving the
abstraction barrier. The states which underlie the various uses of a concept
need not themselves be symbolic states. They are simply whatever it is that
gives rise to the appropriate behaviour. Likewise, the on-line performance
determining network and the more theoreticised network may each be directly
prone to go from certain inputs to appropriate outputs without the benefit of
intervening stages of inference in a quasi-sentential symbolic formalism. To
the extent that we find all this hard to imagine it is surely due to our habit of
producing public linguistic rationales when called upon to justify our actions.
(See Malcolm et al (1989)p.7). but whoever said that the sub-parts of the
brain have to follow the same kind of routine as we (the molar agents) do?
The organisational requirement we endorse simply demands widespread and
semantically-grounded causal integration and some semantically-grounded
relations of judgemental authority. And semantically grounded just means
that were it not for the semantic commonalities between cases, or the
semantic fact that 'mother of a parent' is constitutive of being a grandmother,
then the internal causal links would not be as they are. The philosophical
constraints outlined earlier thus in no way commit us to symbol-using
architectures. Far from it - they are even compatible with the anti-
representationalist approaches advocated by Brooks and others.

VI. CONCLUSIONS: WHERE PHILOSOPHY MEETS COGNITIVE ARCHITECTURE.

Between out and out behaviourism (innards don't matter at all to the correct
identification of something as a locus of intentional states) and the demand
that all true believers deploy a kind of language of thought, lies the grey area

which I have tried to illuminate. It is surely not the case that any kind of internal organisation whatsoever (just so long as it yields a certain pattern of gross activity) is conceptually acceptable in a putative cognizer. Yet neither should the philosopher find herself too soon legislating that promising new computational models (e.g. connectionism) are incompatible with our vision of ourselves as intentional systems. I have tried to show that attention to the question of whether a given architecture has normative depth may provide a constraint of about the right strength. Such depth is absent from a system in which a 'revised' judgement is causally isolated (both in the past and the present) form the mechanism which issued the original judgement. The giant look-up table lacks normative depth in just this sense. Yet advanced connectionist architectures (and certain versions of anti-representational approaches too) should have not trouble meeting the constraint. This is encouraging news. Perhaps here, at last, we have struck the balance we need?

NOTES

  * Thanks to Christopher Peacocke for a very useful discussion of an earlier draft of this paper. Peacocke's rather different development of the role of normative considerations can be found in Peacocke (forthcoming a and b).
  [1]   There is a proviso viz. that the being must be capable of enjoying a conscious mental life. But Clark (1991) depicts that as the limit of our philosophical concern. The present paper argues that to stop there is to ignore a vital dimension (the ability to make normative commitments - see below) which brings philosophical concerns into intimate contact with the kinds of structural issue debated in cognitive science. The treatment I develop here is otherwise (modulo that revision) compatible with the general thrust of the 1991 paper.
  [2]   Notice that this appears to be weaker than the requirement defended in Davies (1991) insofar as it allows that different inner resources may be operative on different occasions in which the same concept is deployed, just as long as the inner resources share relevant causal properties because of commonalities in their contents. Context-sensitive connectionist representations thus meet the demand. Here, then, we depart radically from Davies (1991) which parlays the constraint into an a priori argument for the language of thought hypothesis. (See Fodor, J. (1987)).
  [3]   This is no accident. In the actual universe of limited resources, head-size and processing-time, it is likely that only beings who display such internal organisations will succeed in producing the requisite behaviour. This empirical fact, however, cannot license a conceptual instrumentalism (as Peacocke's example of the Martian marionette (Peacocke (1983)) shows).
  [4]   See e.g. the mobile 6-legged insect described in Brooks (1986) and

the various examples in the survey of work in the 'new robotics paradigm' in Malcolm, Smithers and Hallam (1989). Of course, it may well be that the very radical approaches which eschew all forms of symbolic representation will fail to have the resources to model truly high level cognitive achievements. That is an empirical issue (though not, I suspect, one with as cut and dried an answer as many classicists believe!). Our question, which is quite independent of this, is whether a radically non-symbol using system could at least meet the minimal organisational demands we have suggested.

REFERENCES

Brooks, R.: 1987, Planning is just a way of avoiding figuring out what to do next, M.I.T. Artificial Intelligence Laboratory Working Paper 103.

Churchland, P.: 1989, *The Neurocomputational Perspective: The Nature Of Mind And The Structure Of Science.* Cambridge, Ma: MIT/Bradford Books.

Clark, A.: 1989, *Microcognition: Philosophy Cognitive Science And Parallel Distributed Processing.* Cambridge, Ma: M.I.T. Press, Bradford Books.

Clark, A.: 1991, Radical Ascent, In *Proceedings Of The Aristotelian Society* Supp.Vol. LXV, pp. 211-227, Aristotelian Society

Clark, A. and Karmiloff-Smith, A.: 1993, The cognizer's innards: a psychological and philosophical perspective on the development of thought, *Mind And Language* **8**, pp. 487-519.

Davies, M.: 1991, Concepts, connectionism and the Language of Thought in W. Ramsey, S. Stich and D. Rumelhart, (eds.), *Philosophy And Connectionist Theory*, pp. 229-258. (Hillsdale, NJ: Erlbaum)

Evans, G.: 1982, *The Varieties Of Reference*, Oxford: Oxford University Press.

Fodor, J. and Pylyshyn, Z.: 1988, Connectionism and cognitive architecture. A critical analysis, *Cognition* 28, pp. 3-71.,

Legendre, G., Miyata Y., and Smolensky, P.: 1990a, Harmonic grammar: a formal multi-level connectionist theory of linguistic well-formedness: an application, Technical report Cu-Cs-464-90, University of Colorado at Boulder.

Legendre, G., Miyata Y., and Smolensky, P.: 1990b, Harmonic grammar: a formal multi-level connectionist theory of linguistic well-formedness: theoretical foundations, Technical Report CU- CS-465-90, Department of Computer Science, University of Colorado, Boulder.

Malcolm, C., Smithers, T. and Hallam, J.: 1989, An emerging paradigm in robot architecture, in Edinburgh University Dept. Of Artificial Intelligence Research Paper no. 447. Presented at the Intelligent Autonomous Systems Conference, Amsterdam, 1989.

Mozer, M. and Smolensky, P.: 1989, Using relevance to reduce network size automatically, *Connection Science*, vol. 1, no. 1 , pp. 3-17.

Peacocke, C.: 1983, *Sense And Content: Experience, Thought And Their*

Relations, Oxford: Clarendon Press.

Peacocke, C.: (forthcoming - a), Philosophical and Psychological Theories of Concepts. in Clark, A. and Millikan, P. (eds.) *Proceedings Of The 1990 Turing Colloquium.* (Oxford: Oxford University Press)

Peacocke, C.: 1991, Contents and Norms in a Natural World. in E. Villanueva (ed) *Information, Semantics And Epistemology.* Oxford: Blackwell.

Pinker, S. and. Prince, A.: 1988, On language and connectionism. Analysis of a parallel distributed processing model of language acquisition, *Cognition*, no. 28:, pp. 73-193.

Smolensky, P.: 1991, Connectionism, constituency and the language of thought, in *Meaning And Mind: Fodor And His Critics.*, ed. B. Loewer and G. Rey, Oxford: Blackwell.

School of Cognitive and Computing Sciences
University of Sussex
Brighton BN1 9QH

KEITH LEHRER

SKEPTICISM, LUCID CONTENT AND
THE METAMENTAL LOOP

Cognitive science has concerned itself rather little with knowledge and even with our knowledge of the content of our own thoughts. The focus of inquiry is more often representation and the functional role thereof. But knowledge is more than representation, even representation that something is the case, for we can have a representation and not know that the representation is correct. This paper is concerned with knowledge and the special kind of knowledge that we have of the content of our own thoughts. I call the content of our own thoughts, *lucid content*, and shall offer an explication of it. My primary thesis is that metamental processing and evaluation convert mental representation into knowledge and mental activity into lucid content. My task is to explain how this can be so.

I. KNOWLEDGE AND EVALUATION.

Let me begin with knowledge and some account, however brief and truncated, of what knowledge is.[1] Knowledge goes beyond the mere receiving or representation of information no matter how reliable the source and mechanism of it. If I have no idea whether some information or representation that *p* is correct or not, then, obviously, I do not know that the information or representation that *p* is correct, and, therefore, I do not know that *p*. Moreover, what is required to convert representation into knowledge is important to the study of cognition and cognitive science. The conversion is effected by evaluation of the representation in terms of a background system of the previous acceptances of representations. The evaluation of the incoming information is the task of some central system that has access to the previous acceptance of information. This evaluation provides us with a filter of incoming representation, and knowledge is successful filtering.

The most salient cases in which representation falls short of knowledge are those in which the subject accepts some hypothesis or theory that implies that the circumstances in which the representation arises are

A. Clark et al. (eds.), Philosophy and Cognitive Science, 73–93.
© 1996 *Kluwer Academic Publishers.*

untrustworthy. For example, if the subjects of an experiment are given instructions to the effect that they are being placed in a room in which they will be deceived about the colors of objects by the use of colored lights and other deceptive devices, the fact that they receive information which gets represented as the claim that they see something red does not constitute knowledge that they see something red. The reason is simply that the instructions that they are given put them into a situation where they have no way of telling whether the representation is correct or not. They have, or should have, no idea whether see something red or not Notice that it is not assumed that the instructions are themselves correct or that they do not see something red. The instruction may be incorrect, the situation may be normal, and they may actually see something red. But having accepted the instructions to the effect that the situation is abnormal, they have no way of knowing whether their visual color representations are correct or not.

The foregoing argument is the stuff of which skeptical arguments have been made and are made in philosophical argumentation. Such arguments incorporate a duplication thesis, one to the effect that deceptive circumstances can be conceived in which representations of some kind, perceptual ones for example, will duplicate the representations that arise in what we assume to be normal circumstances, but will be such that we have no way of telling whether our circumstances are the deceptive ones or the normal ones. The correct point of such skeptical arguments is that if we have no idea whether our representations are correct or not, they fall short of knowledge. I do not raise this point to retrace the Cartesian path to the doorstep of paradox. I raise it order to reveal the way in which representations must fit in with our background system, the way they must cohere with it, in order to be converted into knowledge.

II. COHERENCE AND THE METAMENTAL.

What is a background system and how must something cohere with it to yield knowledge? The background system is a system of information consisting of evaluations of lower level representations in terms of the purpose of obtaining truth and avoiding error. When the evaluation is positive, I say that the first order representation is accepted. Acceptance is an amalgamation of a lower level representation and higher order or metamental evaluation of the lower level representation. The notion of acceptance like the notion of the metamental is a relative notion. The evaluation of any representation of any level is at a metamental level relative to the representation evaluated.

If one thinks of some representation as resulting from an input system encapsulated from the background system, then one will think that there are innocent first order representations untainted by metamental evaluation, but innocence in information processing, as in life generally, is soon lost in higher order sophistication. Moreover, it is innocence well lost, for unevaluated information is not the stuff of which knowledge is made. Prior to evaluation we have no idea whether the representation is correct or incorrect, true or false, and, therefore, no idea whether it is correct or misleading information. Lower level creatures lacking the capacity for metamental ascent have no way of ascending from representation to knowledge. The background system consists of stored evaluations or acceptances. It is important to note that the system of acceptances is not just a system of the representations accepted but is a system of the positive evaluations or acceptances of those representations.

III. COHERENCE AND SKEPTICISM

This leaves us with the problem of indicating how coherence with the acceptance system converts representations into knowledge. I propose a two step solution to the problem. The first step may be put by saying that the acceptance system must be adequate to meet skeptical objections concerning the representation. There are, moreover, two kinds of skeptical objections. The first consists of those that simply contradict the representation, and the second consists of more interesting objections that shed doubt on the trustworthiness of the representation without contradicting it. The example considered above in which I accept that I am in deceptive circumstances is a skeptical objection of the second kind, for this does not entail that I do not see something red when I think I do, only that such representation is not trustworthy in the circumstances in which I find myself.

There are also two ways of meeting a skeptical objection in terms of an acceptance system. The objection may simply be beaten by the acceptance system by which I mean that positive evaluation or acceptance of the representation is more reasonable on the basis of the acceptance system than acceptance of the competitor. In ordinary circumstances, in which I accept that I can tell a red thing when I see one, the competitor that I am deceived about the color of the object I see when I accept that I see a red object will be beaten. The positive evaluation of the representation that I see something red is more reasonable in terms of my background system than the skeptical objection that I am deceived about the color of the object I see. One might imagine a dialogue between a claimant, who claims to know, and a skeptic wherein the claimant meets the objection as follows:

> Claimant: I see a red object.
> Skeptic: You are deceived about the color of the object.
> Claimant: It is more reasonable for me to accept that I see a red object than that I am deceived on the basis of my acceptance system. The circumstances are ones in which I can tell a red object when I see one in a trustworthy manner. The light is good, there is no peculiar lighting, my eyes are good, and I am otherwise normal in ways relevant to trustworthy acceptance of what I see.

The second way of meeting a skeptical objection is to neutralize the objection which is necessary when the objection is something that it is quite reasonable to accept, for example, that people are sometimes deceived about the color of objects. In this case, the objection is met by noting that, though people are sometimes deceived, I am not deceived, and, moreover, it is just as reasonable for me to accept that as to accept the skeptical objection alone on the basis of my acceptance system. In this case the dialogue between the claimant and the skeptic runs as follows:

> Claimant: I see a red object.
> Skeptic: People are sometimes deceived about the color of objects.
> Claimant: It is as reasonable for me to accept that people are sometimes deceived about the color of objects and that I am not deceived as to accept the former alone on the basis of my acceptance system. The circumstances in which people are deceived are not present here. The present circumstance are those decribed in my previous reply in which I can tell a red object when I see one in a trustworthy manner.

When a person accepts something and skeptical objections can all be met, that is, either beaten or neutralized in terms of the background system, then,

I shall say that the person is personally justified in accepting the thing in question. Personal justification is a condition of knowledge.

It is important to note that, though the acceptance system of a person must suffice to meet skeptical objections to yield personal justification, a person obviously need not process all such objections. The acceptance system need only contain the activation potential to meet such objections and, in fact, if the person accepts her trustworthiness concerning the representation, then the activation potential results from a course grained and perhaps somewhat dogmatic disposition to respond to objections with the counterassertion that it is more reasonable to accept the representation, for example, that I see something red, then the skeptical objection because I accept that I am trustworthy about the matter. Though the potential objections are many in number, the capacity to respond adequately to them on the basis of the acceptance system may be effected by a psychologically simple strategy. Nevertheless, in the example considered above in which a person accepts that the circumstances are deceptive, the simple strategy for meeting the skeptical objection is blocked because I do not accept that I am trustworthy about the matter.

An epistemologist, and perhaps a cognitive scientist as well, will quickly note that personally justified acceptance will not suffice for knowledge, because what a person is personally justified in accepting might be false and, indeed, the acceptance system on which the justification is based might be rife with error. This point will, perhaps, be of less interest to a cognitive scientist than to a philosopher because personal justification exhausts the internal psychological conditions germane to justification and knowledge. Nevertheless, the need to add some condition to personal justification to convert it to knowledge is important, and what must be added will turn out to have some relevance to disputes in philosophy of mind and cognitive science. The addition is the second step in the conversion alluded to above.

IV. THE TRUTH CONDITION

What must be added is some truth condition pertaining to acceptance. It would, of course, be unrealistic to require that everything contained in the acceptance system of a person be true, but we might add a more modest constraint that for personal justification to convert to knowledge corrections of errors in the acceptance system must not defeat the personal justification.

Thus, if I accept that I am trustworthy in the circumstances in which I accept some perceptual representation when, in fact, I am not trustworthy in the circumstances, then correction of this error in my acceptance system, say by substituting acceptance that I am not trustworthing about such perceptual matters in the circumstances for acceptance that I am trustworthy, my justification will defeat. The technical problems of articulating an adequate account of this notion of defeat are not trivial, and I have spent many bytes over them, but the foregoing remarks will suffice for our purposes.

Our two step account of the conversion of acceptance into knowledge may be put as follows: knowledge is undefeated personal justification. This account yields two consequences worth noting before we turn to applying this account to our knowledge of lucid content and the implications of such knowledge. The first is that the account yields a truth a condition, that is, if I know something I accept, then it must be true. For, if I accept something false, then a correction of that error in my acceptance system would require substituting acceptance of the denial of it for the acceptance of it, and the result would be the defeat of my personal justification. The second is that the account contains an internal prophylactic against skeptical argumentation. Consider a skeptical hypothesis intended to undermine a knowledge claim. If a person accepts things that refute that hypothesis and imply that she is trustworthy in the matter, then she may be personally justified in accepting what she does contrary to the skeptical hypothesis. This, of course, does not guarantee that the person has knowledge, for, if the skeptical hypothesis is correct, then what the person accepts to refute the skeptical hypothesis will be false and, consequently, the personal justification will be defeated. But the mere presentation of the skeptical hypothesis and our failure to prove it false does not sustain the skeptical conclusion. If we are trustworthy as we take ourselves to be, then our personal justification will go undefeated and acceptance will become converted into knowledge.

V. METAKNOWLEDGE AND THE METAMENTAL LOOP

Knowledge results from the positive evaluation of the trustworthiness of our information and from the correctness of that evaluation. In this way, knowledge involves the metamental and is impossible without it. Knowledge is, therefore, metaknowledge. This account might appear to lead to a traditional regress. If we need to evaluate the trustworthiness of the information we receive, do we not also need to evaluate the trustworthiness of the information that the information we receive is trustworthy? And will that not create the need for ever higher levels of evaluation *ad infinitum*?

The solution to the problem depends on distinguishing various claims about our trustworthiness. Consider the following claims:

> S.   The source of my information is trustworthy.
> O.   The operations of my mind on the information
>   that I receive are trustworthy.
> T.   I am trustworthy in what I accept.

I accept S and O. If T is true, then I am trustworthy in accepting S and O. On the other hand, if S and O are true, that supports T. What is most interesting about T, however, is the way in which acceptance of T supports itself. If I accept T, then T tells me that I am trustworthy in accepting T, because T is one of the things I accept. In this way, T vouches for itself. This metamental loop avoids the regress mentioned above. Acceptance of T is, consequently, a kind of keystone state in our acceptance system. It is that component of the acceptance system that holds the system together to yield personal justification and, if T is true, to convert this into undefeated justification and knowledge. The fundamental reply to the skeptic is a metamental loop.

VI. LUCID CONTENT: A DEFENSE

This reply to the skeptic is, moreover, of relevance to disputes about lucid content to which we now turn. One sort of knowledge that we have is knowledge about the content of our own mental states, what I have called lucid content. Some of our mental states may be epistemologically opaque in the sense that we are ignorant of their existence until we engage in scientific investigation. Other mental states are, however, epistemologically transparent in the sense that we know of their existence prior to any such investigation. The latter include our present thoughts and the content or intentional objects of those thoughts.

What I wish to argue is that lucid content should be taken seriously as part of the data base for cognitive science, that is, that it should be treated as evidence on a par with external perception.[2]   There have been some reasons that have led philosophers and others to doubt that such knowledge should be taken seriously. It has been noted that judgements about our mental states may be in error, that is, that such judgements are fallible.[3] Fallibility is, however, a ubiquitous epistemic condition and, therefore, not a good reason for rejecting any knowledge claim. We can be both fallible and trustworthy, and trustworthiness must suffice for knowledge.

Another reason that has led the cognitive scientist to take on the mantle of skepticism is based on the plausible scientific maxim that the data base for science should be intersubjectively observable. It is then assumed that what is intersubjectively observable is behavior, and it is concluded, therefore, that the data base for science should be behavior.[4]

The reply to this line of thought depends on recognizing that the edifice of scientific knowledge depends on the communication of knowledge and would be impossible without it. Consequently, if we have knowledge to communicate about our own mental states, knowledge expressed in behavioral reports to be sure, that knowledge should constitute part of the data base. Our knowledge of lucid content is something we can communicate and, therefore, should constitute part of the data base. To many of you this will hardly seem necessary to argue for or about, but it has, of course, been a bone of methodological contention.

VII. KNOWLEDGE OF LUCID CONTENT

Application of the analysis of knowledge proposed above reveals our knowledge of lucid content. Consider an example. If I turn my thoughts to what I see, I think that I see a monitor. That is the lucid content of this thought, that I see a monitor. I accept that the content of my thought is just that, that I see a monitor, because I am conscious of that thought.

How it does so, we shall soon consider, but now let us focus on my acceptance that the content of my thought, what I am thinking, is that I see a monitor. Is this acceptance and knowledge appropriate to the data base of science or is it just an error of folk psychology, like demons in wooded caves and spirits in running brooks? Is the acceptance of the lucid content of a thought an error? That is a skeptical doubt. How should the doubt run? It might run with the swiftness of the eliminative materialist. The only thing that is intersubjectively observable is matter, and it is only matter, in this case, brain matter of neural assemblies that we posit in a scientific theory of the mind. When a person accepts that he is thinking that he sees a monitor, what is really going on is that information is received from the monitor by the senses resulting in neural activation in the neural assemblies of the brain. Putting the point crudely, what the person accepts as thinking that he sees a monitor is an error, for all that is going on is vector multiplication in the brain.

Schematically put, the issue runs as follows in a dialogue.

> Claimant: I accept that I am thinking that I see a monitor.
> Skeptic: You are not thinking that you see a monitor for all that is going on is vector multiplication in your brain.

Can I meet this skeptical objection? The reply seems pretty straightforward.

> Claimant: I am conscious of my thinking that I see a monitor, and consciousness of the content of my thoughts is a trustworthy source of information.

This answer is an appeal to my background system. Now we might imagine the skeptic replying as skeptics are wont to do

> Skeptic: Consciousness, whatever that is, leads us to make errors and, therefore, is not a trustworthy source of information.

The reply of the claimant is simplicity itself.

> Claimant: All of our sources of information sometimes lead us to make errors, but that does not show them to be untrustworthy. Moreover, I accept that consciousness is a trustworthy, though fallible, source of information, and I am trustworthy in what I accept.

The last reply of the claimant is intended, of course, to forestall further objections. What the claimant says suffices to personally justify him in accepting what he does about the lucid content of his thought in terms of his acceptance system, in this case, my acceptance system. If, moreover, the claimant's claims are correct then the personal justification will, assuming there are no other objections that cannot be met, convert into knowledge. Simply put, the skeptical objection is met by being beaten. It is more reasonable for the claimant to accept what he does than to accept what the skeptic alleges.

If this reply seems doubtful to you because you are inclined to distrust consciousness as a source of information, you may be unable to assume the position of the claimant. Suppose, however, that you are not inclined to be a skeptic about perception, and, though you have doubts about whether consciousness yields knowledge of lucid content, you do not have such doubts about whether perception yields knowledge of the monitor. You will then find that you are involved in an exactly analogous dialogue with a skeptic who maintains that you do not perceive a monitor at all, that there is only vector multiplication in the brain. The dialogue will proceed as follows.

> Claimant: I accept that I see a monitor.
> Skeptic: You are not seeing a monitor for all that is going on is vector multiplication in your brain.
> Claimant: I perceive a monitor and perception of objects is a trustworthy source of information.
> Skeptic: Perception, whatever that is, leads us to make errors and, therefore, is not a trustworthy source of information.
> Claimant: All of our sources of information sometimes lead us to make errors, but that does not show them to be untrustworthy. Moreover, I accept that perception is a trustworthy, though fallible, source of information, and I am trustworthy in what I accept.

In short, the defense of perception of external objects against skepticism is exactly parallel to the the defense of consciousness of lucid content against skepticism. Thus, if perception is worthy of scientific trust, so is consciousness.

It should be noticed that a more modest skeptic cannot be beaten but may be neutralized. Suppose our materialistic skeptic puts forth a more modest hypothesis, one compatible with token-token identity theory, for example, the hypothesis that what is going on in the brain is neural activation. So the dialogue goes like this.

> Claimant: I accept that I am thinking that I see a
> monitor.
> Skeptic: What is going on in the brain is neural
> activation.

This skeptical hypothesis cannot be beaten, for it is something that it is highly reasonable to accept, but it may be neutralized. For the claimant may reply as follows:

> Claimant: I accept both that I am thinking that I see a
> monitor and that what is going on in the brain is
> neural activation. It is just as reasonable for me to
> accept both as to accept one, and the combination
> does not compete with my original claim.

The claimant may have more than one explanation for why the acceptance of both claims is as reasonable as the acceptance of one, but the most standard explanation would be in terms of the token-token identity of the event of thought and the event of neural activation. This explanation would have the consequence, however, that the event of neural activation has lucid content, indeed, the content that I see a monitor, because the thought event has that lucid content. We would then be left with the problem of attributing intentionality to events of neural activation.

VIII. INTENTIONALITY AND MODEST MATERIALISM

Some philosophers have considered that attribution of intentionality, of lucid content, to neural activation implausible unless the attribution can be reduced to attribution of known properties of matter.[5] If the reduction is impossible, they would conclude that the attribution of intentionality must be erroneous. But this conclusion is badly drawn, and we should not allow ourselves to be drawn to it. Our knowledge of the lucid contents of our thoughts, of what they are about, of their intentional objects, however one cares to put it, is a known fact for science to explain. It is not the business of philosophy or science to deny known facts to simplify our theories or fit our reductionistic methodologies. I subscribe to materialism of a sort, what I prefer to call modest materialism. Modest materialism tells us the the only things that exist are material objects and their properties, but, in modesty, it refuses to prejudge what the properties of matter must be.[6] Following his premises, a modest materialist who knows that his thoughts have lucid content will attribute lucid content to the body, and, most plausibly, to neural activation in the brain. If, as some philosophers have held, the attribution of intentionality to the brain mentalizes the body, then the mentalization of the body is a necessary condition of accounting for our knowledge of it. We know that our thoughts have lucid content.

IX. THE PROBLEM OF LUCID CONTENT

The problem that now confronts us is to account for our knowledge of lucid content. Standard accounts of wide and narrow content shroud our knowledge of the content of our thoughts in mystery.[7] Let us begin with the assumption that thought is mental tokening, though, in fact, other accounts of thought as activation will do quite as well as I shall argue later. Standard accounts of wide content explicate wide content in terms of causal etiology or covariation with the tokening. These accounts are difficult to render plausible because of the multiplicity of causes and reasons we have for tokening a sentence, only one of which is accurate description, but the problem that arises from equating lucid content with wide content does not depend on details. The problem is the ease and immediacy with which we know the contents of some of our thoughts, their lucid contents, is incompatible with the assumption that lucid content is wide content. We have no way of knowing etiology or covariation with armchair ease and immediacy.

Narrow content might be thought to fare better, but narrow content construed as the functional role of a token in a complex network of nomological relations to other tokens cannot be easily known either. We have no way of knowing the functional role with ease and immediacy. Some internal state that in some way accompanies narrow content or wide content, for that matter, might, of course, be known immediately, but to know that the internal state accompanied wide or narrow content would be necessary for our knowledge of lucid content to be knowledge of narrow content. This knowledge would, again, lack the ease and immediacy of our knowledge of lucid content.

X. CONSCIOUSNESS AND THE METAMENTAL LOOP

Suppose, then, that a thought is mental tokening, tokening, that is, of some mental sentence. How do we know the lucid content of the sentence tokened? The answer is that consciousness produces a nexus of metamental ascent and descent, quotation and disquotation, that, combined with our understanding of the sentence, yields our knowledge of lucid content. Consciousness creates a kind of metamental loop from quotation of the token back on itself as an exemplar of a token having a certain kind of function or role. Let me explain by appeal to another loop. Consider the sentence

> S.   This sentence refers to itself.

How do we know that this is true? The sentence refers to itself and tells us that it is true. Understanding the sentence suffices for knowing that the sentences is true because in referring to itself it becomes an exemplar of the class of things that refer to themselves to which it belongs.

Now consider the sentence of thought

> M.   I see a monitor

the tokening of which is involved in my thought. How does consciousness give us knowledge of the content of this thought involving the token M? Consciousness gives us metamental ascent by quoting the sentence and metamental descent by disquoting the sentence to yield

> LC.   "I see a monitor" has the content that I see a
>   monitor.

The yield is, of course, not very informative, but, if I understand LC, then it tells me the content of my thought. The trick of consciousness is to effect a loop from the token back onto itself as a token representing an understood class of tokens. The final four words in LC reuse the token to refer to a class of tokens of which it is a member and identify the class by exhibiting itself as a member of the class.

It is very important to an understanding of the role of consciousness in our knowledge of lucid content not to confuse the original tokening of the sentence with the quotation of the token. Even the sentences

   M. I see a monitor

and

   MT. The sentence "I see a monitor" is true

do not have the same content or meaning. Sentence M does not entail that there are any sentences, it only says that I see a monitor, while sentence MT entails the existence of the sentence. Yielding LC, consciousness adds to our understanding by means of the metamental loop of the token onto itself.

   It is now simple enough to understand how consciousness gives us knowledge of the lucid contents of our own thoughts. Consciousness reveals what a sentence is about by the loop of metamental ascent and descent using the sentence itself to reveal what the sentence is about. Moreover, in revealing what the sentence is about, it reveals itself as the revealer of this fact. Consciousness reveals itself in the same way that light does. Light in revealing the illuminated object reveals itself at the same time. Thus our knowledge of the lucid content of our thoughts is at the same time knowledge of our consciousness of that content. Consciousness is consciousness of itself at the same that it is consciousness of content.

   We may remove the mysterious of supposing that some token can be made into an exemplar or, in a sense, a symbol of a class of things to which it itself belongs by considering some other examples. A token of a song can stand for a class of tokens of a given kind or type.[8] A token of a song may be sung as an exemplar of the class of tokens of a song. So, for example, if I tell you that my favorite pop rock song is the *Shoop Shoop Song* sung by Cher, and you want to know what the song is, I might sing it for you, or, much better, get Cher to sing it for you. The token produced serves as an an exemplar of the tokens of the song. If you are inclined to think of the song as the class of token singings, then the token sung will stand for the class at the same time that it is a member of the class. If you are inclined to think of the song more platonistically as songs of certain kind, then the token sung will stand for tokens of that kind. There is, however, a loop of reference from the token to class of tokens of which it is a member.

XI. KNOWLEDGE OF MENTAL STATES AND THE METAMENTAL LOOP

The nexus of consciousness supplying a loop from a token back on itself as an exemplar of a kind of token to give us knowledge of lucid content has more general implications. My speculation is that the metamental loop of consciousness converts mental states into symbols in a way that accounts for our immediate knowledge of many of our mental states, not just our intentional states. Consider our knowledge of sensations which are alleged not to be intentional. The metamental loop of consciousnes yields knowledge of those states, of a pain, for example, by making a symbol out of the pain. Just as we take the token of the song to stand for other tokens of the same kind and thus know what the *Shoop Shoop Song* is and that what is tokened is the *Shoop Shoop Song*, so we take the token of pain to stand for other tokens of the same type and know both what pain is and that what is tokened is pain. Our knowledge that we are in pain is a result of the metamental loop of consciousnes using the pain as an exemplar or symbol of pain.

This feature explains the traditional doctrine of our incorrigible knowledge of our mental states. Using the state as an exemplar to stand for a class of states, we would seem invulnerable to error in believing that the state belongs to class of which it is an exemplar. Unfortunately, consciousness, like perception and other faculties, can misfunction in the production of the nexus and lead us into error. We need the assumption, noted above, that consciousness as the source of our information is trustworthy, which introduces coherence into our knowledge. Moreover, a particular pain, though it may represent pains generally, is a poor symbol for the purposes of memory and communication. Consequently, other symbols are needed, like the word "pain" which are subject to all the hazards of erroneous application. To store and communcate our knowledge of our mental states we must map them into more conventional symbols than the metamental loop consciousness provides. It may, however, provide us with some relief from the arrows of skepticism to reflect that the metamental loop of consciousness can provide us with knowledge of the lucid content of thoughts and of the existence of other mental states, though, it goes without saying, only those of which we are conscious.

XII. LUCID CONTENT WITHOUT MENTAL SENTENCES

Consideration of the way in which a metamental loop can provide us with representation and knowledge of mental states will allow us to dispense with the assumption that all our knowledge of the lucid contents of our thoughts depends on mental sentence tokens. The thesis of one version of the computational theory of mind ascribes mental tokens in the language of thought to us as explanatory entities. It is possible, however, that the advantages of such a theory might be obtained without the postulation of such entities, as I have argued elsewhere, provided that operations of thought have sufficient logical and semantic complexity. Of course, there are sentences that occur in thought, the sentences of conventional languages, English, Spanish, Esperanto, and so forth. We use such sentences to articulate our thoughts to ourselves or others. The metamental loop operates on these whose existence belongs to the empirical database for cognitive science. The postulation of an innate language of thought used to interpret these sentences and other expressions of conventional languages goes beyond the database to theory and the postulation of theoretical entities. That is not a criticism of such a strategy, but I prefer not to tie the account I am presenting to the postulated tokens of theory of the language of thought.[9]

Fortunately, it is not difficult to extend the account of how a metamental loop gives us knowledge and representation of mental operations to yield an account of lucid content without postulating the existence of mental tokens in the language of thought. Suppose that there are thought processes having lucid content without any mental token having that content. For example, suppose that there is some operation, O, of the mind that has the content that I am thinking that I am in pain. That operation, O, can be used to represent itself and other operations of the same kind in the way the pain or the song can be used to represent itself and other things of the same kind. The metamental loop can start from any operation of the mind and by means of the loop of metamental ascent and descent yield a representation of the operation. Looping can, in principle, convert any operation of thought of which we are conscious into a symbol and, by so doing, yield immediate knowledge of the lucid content of the thought.

XIII. NEURAL ACTIVIATION: AN INTEGRATIVE LOOP

I have argued for the importance of metamental ascent for our knowledge of content and have tried to explain how it works. In conclusion, I want to close the metamental loop by coming back to the body and to neural activation. You might wonder by this time whether the notions of metamental ascent and the metamental loop have so far removed us from neural activation that, if you follow me where my conclusions draw us, you will have eliminated the possibility that the brain realizes such mental activity. In fact, I want to suggest there is a loop in neural activation. My argument is simple and mathematical, though left here without neurological detail. These are the assumptions of my argument. It is necessary for the brain to average incoming input across neurons or neural groups, and the brain must find the appropriate set of weights for averaging. Neural activation realizes weighted averaging. The weights are a set of numbers that are nonnegative and sum to one. It is, moreover, important to notice that averaging of input is one standard model of aggregating data, especially conflicting data. The weights that are used for averaging are often the set of prior probabilities of some set of disjoint and exhaustive alternative evidence statements.

XIV. PROBABILITY AGGREGATION AND CONSENSUS

For example, when there are various probabilities that might be assigned to a hypothesis, H, on the basis of various possible statements of evidence, E1, E2 and so forth to En, which are disjoint and exhaustive, the probability of H is computed from the following fundamental theorem of Bayesian probability:

$$p(H) = p(H/E1)p(E1) + p(H/E2)p(E2) + ... + p(H/En)p(En)$$

Another example is the Jeffrey formula for computing a new probability, p_N, of H as the result of assigning new probabilities to the evidence sets E1, E2 and so forth to En by appealing to the old conditional probabilities, p_0, of H on the evidence sets as follows:

$$p_N(H) = p_0(H/E1)p_N(E1) + p_0(H/E2)p_N(E2) + + p_0(H/En)p_N(En)^{[10]}$$

In these cases, the weights used to average the conditional probabilities are prior probabilities.

The weights used in aggregation need not be prior probabilities, however. One example of aggregation by weighted averaging that does not use prior probabilities is the aggregation of conflicting probabilities that different people assign to the same hypothesis at the same time[11]. Thus, suppose that a group of people have probabilities assignments, where $p_i^0(H)$ is the probability assigned to H by person i at a given stage 0 , and each person assigns some weight to every other person, where w_{ij} is the weight person i assigns to person j as a measure of the comparative trustworthiness of j as an evaluator of the probability, such that the weights i assigns to members of the group are nonnegative and sum to one. Person i might then improve her probability assignment for H on the basis of her evaluations of the trustworthiness of members of the group by computing a stage 1 probability, p_i^1 , as a weighted average of the probabilities of the stage 0 probabilities as follows:

$$p_i^1(H) = p_1^0(H)w_{i1} + p_2^0(H)w_{i2} + ... + p_n^0(H)w_{in}.$$

This method of aggregation has the peculiarity that if iterated from stage S to stage S+1 according to the more general formula

$$p_i^{S+1}(H) = p_1^S(H)w_{i1} + p_2^S(H)w_{i2} + ... + p_n^S(H)w_{in}$$

will yield the result that the probabilities of members of the group will converge toward a consensual probability $p_C(H)$ when there is a chain of positive respect connecting members of the group. Mathematically considered, the process of iterated averaging is equivalent to finding a consensual weight, w_j, for each member of the group and computing the weighted average of the initial stage 0 probabilities to find the consensual probability as follows:

$$C. \ p_C(H) = p_1^0(H)w_1 + p_2^0(H)w_2 + ... + p_n^0(H)w_n.$$

It is easy enough to understand the mathematics of consensus in the two membered case. If I give you a weight of .1 and myself a weight of .9 and average while you reciprocate, I shall move .1 toward your probability assignment as you move .1 toward mine moving from stage 0 to stage 1. If we repeat the process we will converge toward the probability assignment that would have resulted from each of us assigning .5 to each other originally and using the weights we assign to compute a weighted average of the original probabilities we assigned. We would, of course, arrive at a common consensual probability.

XV. THE VECTOR LOOP

Before returning to the brain, it is useful to notice that weights in formula C . are unique and have a very special property that makes it easy to compute those weights without going through the process of iterated averaging, though the process of iterated averaging will find those weights. Consider the weights w_{1j}, w_{2j}, and so forth to w_{nj} that the various members of the group assign to member j. Now suppose that we want to average these diverse weights assigned to j to find the consensual weight to be assigned to j, w_j. What weights should we use to average those diverse weights assigned by members to j to obtain the consensual weight?

It turns out that there is a unique set of weights, the fixed point vector, that will yield the consensual weight for j that iterated averaging finds, namely, the set of consensual weights themselves. So the answer to our question is contained in the following formula:

$$L. \; w_j = w_{1j}w_1 + w_{2j}w_2 + ... + w_{jj}w_j + ... + w_{nj}w_n.$$

Inspection of the formula reveals a loop. We must use the consensual weight for j to compute the weight for j, or, put another way, the consensual weights for members of the group resulting from iterated aggregation are just those weights that yield themselves back when used to average to diverse weights assigned to individuals by members of the group.

XVI. THE NEURAL LOOP

The foregoing model of weighted averaging with a loop may be applied to the brain. Suppose, as has been proposed, that neural activation is weighted averaging of neural input across the neurons. Start with the model of aggregation of probabilities of different people but replace the initial probabilities with initial neural input to a neural unit (a single neuron or some connected group) and consider the weights as modifying the level of activation of the neural units. Finally, suppose that integration of the initial neural input is accomplished by weighted averaging. The task of a neural network would be to find the appropriate set of weights by iterated averaging or some approximation thereof to achieve the integration of the input. The appropriate set of weights to produce integrative equilibrium, integrative weights, would be the fixed point vector, that is, the set of weights achieving equilibrium by yielding themselves back in the process of aggregation.

Thus, if there are weights w_{1j}, w_{2j} and so forth to w_{nj}, that might be used to average input from neural unit j, then formula L. above shows us that there is a fixed point vector containing an integrative weight for each neural unit. The fixed point vector, itself found by iterated averaging, finds an equilibrium integration of neural input. It is, moreover, precisely that unique vector that yields itself back producing equilibrium in the process of aggregating neural input.

The foregoing proposal is, of course, highly schematic as well as speculative. It does, however, make an empirical prediction. The prediction is that the neural unit would need to receive considerable neural feedback or backward projection from other neural units in the network in order to find the integrative weights. Assuming that there is no localized center in the brain for computing the fixed point vector, which surely there is not, the integrative weights must be found by feedback from neural units in the neural network altering the activation levels of the neural units. The interaction of the neural units can find the integrative weights or an approximation thereof without any localized center for accomplishing this task only if there is considerable backward projection in the neurons in comparison to the forward projection of input activation. Backward projection could effect the integrative loop.

My concluding conjecture is that an activation loop is a possible neural realization of the loop of metamental activity. Of course, this is only a possibility, though I think it is a hypothesis worth exploring given the unique role of fixed point vectors in vector aggregation. The possibility shows, however, that there is nothing in the notion of a metamental loop that could not be modeled within a vector activation theory of neural activity. I sought to mentalize the body with intentionality. I sought to metamentalize the mentalized body with knowledge and the metamental loop. I end with a conjecture about how to materialize the metamentalized body in neural vectors with a fixed point vector yielding itself. I hope you enjoyed the loop.

NOTES

[1] I have developed the theory in detail in *Theory of Knowledge*, (Boulder: Westview Press, 1990), chapters 6, 7, and 8, and more briefly in "Metaknowledge," *Synthese* **74**, 1988.

[2] Cf. Lynne Rudder Baker, *Saving Belief*, Princeton University Press, Princeton, 1987.

[3] Cf. Stephen P. Stitch, *From Folk Psychology to Cognitive Science*, MIT/Bradford Press, Cambridge, 1983, and Paul M. Churchland, "Eliminative Materialism and Propositional Attitudes," *Journal of Philosophy*, 78(1981).

[4] Ibid.

[5] Cf. Paul M. Churchland, "Reduction, Qualia, and the Direct Introspection of Brain States," *Journal of Philosophy*, **82** (1985).

[6] Cf. John R. Searle, *Intentionality: An Essay in Philosophy of Mind*, Cambridge University Press, Cambridge, 1983.

[7] Cf. Jerry A. Fodor, *Psychosemantics*, MIT/Bradford Books, Cambridge, 1987, also, "Cognitive Science and the Twin-Earth Problem," *Notre Dame Journal of Formal Logic*, **23** (1982), 98-118.

[8] Cf. Nelson Goodman, *Languages of Art*, Bobbs-Merrill Pub. Co, Indianapolis, 1968.

[9] Cf. Jerry A. Fodor, *The Language of Thought*, Thomas Y. Crowell, New York, 1975.

[10] Richard C. Jeffrey, *The Logic of Decision*, McGraw Hill, New York, 1965.

[11] Cf. Keith Lehrer and Carl Wagner, *Rational Consensus in Science and Society: A Philosophical and Mathematical Study*, Reidel, Dordrecht, 1981.

Department of Philosophy
University of Arizona
Tucson, Arizona 85721

JOHN PERRY

EVADING THE SLINGSHOT[1]

The topic of this essay is "the slingshot", a short argument that purports to
show that sentences[2] designate (stand for, refer to) truth values. Versions of
this argument have been used by Frege,[3] Church,[4] Quine[5] and Davidson,[6]
thus it is historically important, even if it immediately strikes one as fishy.
The argument turns on two principles, which I call substitution and
redistribution. In "Semantic Innocence and Uncompromising Situations",[7]
Jon Barwise and I rejected both principles, as part of our attempt to dismantle
the slingshot and defend the view that sentences stand for complexes of
objects and properties rather than truth values. In his book *An Essay on
Facts*,[8] Ken Olson maintains that our treatment turns on the structuralist
conception of facts, and that this conception leads either to a block universe
of co-implicating facts, or bare particulars. I'll first review the case against
the slingshot, and then consider the issues Olson raises.

I. DO SENTENCES DESIGNATE?

As a preliminary we need to consider the very idea that sentences designate
anything at all. We ordinarily talk about what terms refer to, stand for, or
designate, but do not use these locutions with respect to sentences. Why
should we? Because we want to systematically connect the designation of
complex expressions with the designations of their parts. Many complex
expressions have sentences as parts; to extends our principles of designation
to such expressions, we need to accord designation to sentences.
  Consider (1) and (2).

(1)   The total number of votes Bush received
(2)   The total number of votes Dukakis received

It is natural to say that (1) and (2) designate numbers. (1) and (2) designate
the numbers they designate, in part because "Bush" and "Dukakis" designate
the persons they designate. This suggests the principle that the object an

95

A. *Clark et al. (eds.), Philosophy and Cognitive Science*, 95–114.
© 1996 *John Perry.*

expression designates helps to determine the object larger expressions of which it is a part designate. An account of expressions of the common form of (1) and (2) would make this dependence clear:

> Des("The total of number of votes α received") = The total number of votes Des(α) received.

Des is a function from an expression to its designation. We see in the principle how the designation of the part, Des(α) on the right side, contributes to the determination of the designation of the whole, the left hand side. Such principles identify two roles for designating expressions: being a part that contributes and being a larger whole that receives a contribution. Sentences can play both roles. If we can identify some factor connected with sentences that is systematically determined by what their parts designate, and is systematically determines that same factor with respect to the larger sentences of which they are parts, it will not be stretching things too far to call that factor what the sentence designates.

Sentences have factors associated with them that are systematically determined by what their parts designate, and sentences contribute something to the determination of this factor for the larger expressions of which they are parts. (3) and (4)

> (3)   Bush won.
> (4)   Dukakis won.

have different *truth conditions* and different *truth values*, because there is a difference in what "Bush" and "Dukakis" designate. These dependencies are reflected in this principle:

> "α won" is true iff Des(α) won

(3) and (4) are parts of (5) and (6):

> (5)   It is not the case that Bush won.
> (6)   It is not the case that Dukakis won.

The truth conditions and truth values of (5) and (6) clearly depend on the truth conditions and truth values of (3) and (4). (5) is false because (3) is true; (6) is true because (4) is false.

In the case of (5) and (6), we could take the designation of sentences to simply be truth values. But for a wide variety of cases, truth values do not seem to work as the designata of sentences. (7) and (8) are both true, while (9) is true and (10) false, so (7) and (8) must be contributing something besides their truth values to (9) and (10):

(7) $2 + 2 = 4$
(8)   Konigsburg is in Russia
(9)   Necessarily $2 + 2 = 4$
(10)   Necessarily Konigsburg is in Russia

Here it seems that the difference in truth *conditions* between (7) and (8) accounts for the difference in truth value of (8) and (9). The conditions of the truth of (7) are met no matter what, while those of (8) are quite contingent. So we might be inclined to think that at least for a wide range of cases we should accept (A)

(A)   A sentence designates its truth conditions.

I'll present the slingshot as an attempt to show that (A) leads, in spite of its motivation, to truth values as the designata of sentences: even if we want truth conditions, we end up with truth values. Since we know that truth values won't work for cases like (7)--(10), this is an unwelcome result.

Intuitively, whether the truth conditions of a sentence are met or not will come down to which properties objects have and which relations they stand in. Two sentences that will be true if just the same objects have just the same properties and stand in just the same relations will have the same truth conditions. If sentences are the same in this way, its does not seem like it should matter how those conditions get presented or built up. These considerations appear to support two further principles:

(B)   Substitution of one co-designating term for another does not affect the truth conditions of a sentence.

(C)   Sentences whose truth requires the same objects to
           have the same properties have the same truth
           conditions, even if they differ in syntactic
           structure, and so construct requirements in
           different ways.

(A), (B) and (C) seem to guarantee two principles of "designation-preservation" for sentences.

Substitution: (From (A) and (B)) Substitution of one co-
           designating term for another does not effect what a
           sentence designates.
Redistribution: (From (A) and (C)) Rearrangement of the
           parts of a sentence does not effect what it designates,
           as long as the truth conditions remain the same.

The slingshot starts with a sentence, and then moves, by a series of substitution and redistribution steps to a completely different sentence. Since one gets from one sentence to the other by these steps, they must designate the same thing. But the only thing the sentences have in common are their truth values, so this must be what they designate. We'll look at two versions.

II. TWO SLINGSHOTS

The first version is inspired by Church (Church, 1956):

C1   Scott is [the author of *Ivanhoe*].
C2   Scott is [the author of 29 *Waverley* novels
          altogether].
C3   29 is [the number of Waverley novels Scott wrote
          altogether].
C4   29 is [the number of counties in Utah].

The steps from C1 to C2 and C3 to C4 are substitution steps. The bracketed expressions in C1 and C2 designate the same object, Scott. The bracketed expressions in C3 and C4 designate the same object, the number 29. The step from C2 to C3 is a redistribution step. Since both substitution and redistribution preserve what is designated, C1 must designate the same thing

C4 does. But then it seems like what is designated must just be truth values, for what else do C1 and C4 have in common?

The argument looks like a big trick. Let's call any property, relation or object designated by a simple expression in a sentence part of that sentence's *subject matter*. The step from C1 to C2 changes the subject matter; some of C1's subject matter is lost, and some new subject matter is introduced. In C3 the subject matter is redistributed, and in C4 substitution introduces new subject matter again, while jettisoning Scott, the last vestige of the original subject matter from C1 along with the novels introduced in C2.

Olson reconstructs a version the slingshot from Gödel's discussion of Russell. This argument looks too formal to contain a trick. One needs to assume that every sentence has an equivalent of the form $\kappa(\alpha)$, and that for any two objects there is some true sentence of the form $\pi(\alpha,\beta)$ about their relationship. Let S and T be any two true sentences whatsoever, and $\phi(a)$ and $\psi(b)$ be their equivalents by the first assumption and $\pi(a,b)$ a true sentence by the second assumption. Then, if the first sentence in this series designates a certain object the rest should designate it also:

G1. S
G2. $\phi(a)$ (First Assumption)
G3. $a = \iota x \lfloor \phi(x)\ \&\ x = a]$ (Redistribution)
G4. $a = \iota x\ [\pi(x,b)\ \&\ x = a]$ (Substitution,
                                Second Assumption)

G5. $\pi(a,b)$ (Redistribution A,B)
G6. $b = \iota x\ [\pi(a,x)\ \&\ x = b]$ (Redistribution)
G7. $b = \iota x\ [\psi(x)\ \&\ x = b]$ (Substitution, First Assumption)
G8. $\psi(b)$ (Redistribution)
G9. T   (First Assumption)

We need to emphasize that steps G1-G9 do not represent an *inference* from G1 to G9. We started with the assumption that G1 and G9 were both true. Each step represents a different sentence that can be seen to designate the same thing as the preceding one. The citations on the right do not refer to principles of inference, but to our principles of preservation of sentence designation. So the claim is not that G5 follows from G4, but that, given various facts about the world, including G5 itself, they designate the same thing.

In spite of its formal appearance, this argument turns on the same trick as Church's. The relation π is not part of the subject matter of G3, but is part of the subject matter of G4, while ϕ is part of the subject matter of G3 but not of G4. ϕ gets smuggled out and π smuggled in via the substitution of one description for another. The way the argument works is that the substitution moves changes the subject matter of the descriptions, while the redistribution moves push subject matter back and forth between the descriptions and the predicates. By the time we reach G8, the subject matter has changed completely.

III. TRUTH CONDITIONS AND SUBSTITUTION

As Gödel notes, Russell's theory of descriptions allows him to evade the slingshot. On Russell's theory descriptions are not part of the primitive notation at all. G3 and G4 seem to put the same condition (being identical with a) on the same objects (the ones designated by the descriptions). But on Russell's theory, the descriptions do not really designate anything because they are not really there. If we look at the primitive notation, we will be under no illusions about this:

G3'. Ex[(ϕ(x) & x = a) & Ay (ϕ(y) & y = a --> y = x) & x = a]
G4'. Ex [(π(x,b) & x = a) & Ay(π(y,b) & y = a --> y=x) & x = a]

The Substitution Principle does not get us from G3' to G4' because it does not apply, since the descriptions that are substituted do not occur.

It is not necessary to adopt Russell's theory to avoid the substitution principle, however. It's only necessary to think carefully about truth conditions. If one thinks of designata of sentences as complexes of properties and objects, as Barwise and I were doing, there is an obvious distinction to be made. Consider C1 and C2. Which properties, relations and objects are involved in the truth conditions of C1? Is it Scott twice over and identity? Or Scott, identity, authorship, and the novel *Ivanhoe*? If we take C1 and C2 the first way, they can be thought of as having the same truth conditions. But if we take them the second way, they do not. A condition of C1's truth is that Scott wrote *Ivanhoe*, while this is not a condition of C2's. The situation will be a bit clearer if we shift examples. Consider the following two sentences:

(11)   The author of *Tom Sawyer* grew up in Missouri
(12)   The author of *Huckleberry Finn* grew up in
            Missouri

Do (11) and (12) have the same truth conditions? From one point of view, we might say that they do. The same fact, that Mark Twain grew up in Missouri, makes each of them true. From another point of view, it seems that they do not. For (11) to be true, someone needs to have both written *Tom Sawyer* and to have grown up in Missouri. But this could be true, while (12) was false, and vice versa.

Barwise and I said that there were two ways of building up facts from sentences like (11) and (12), depending on whether one took the descriptions as "value-loaded" or "value-free". The value-loaded interpretations are the same, complexes of Twain, Missouri and the relation of growing up in; the value-free interpretations are different. The latter each involve authorship and a novel, rather than the author.

If we take the descriptions in Gödel's slingshot as the value-free, then the slingshot is blocked at the substitution steps. We can take the designata of descriptions to be complexes of objects and properties, and the designata of sentences to be facts or states of affairs built up out of these.

The situation is not so clear when we take the descriptions to be value-loaded. What then should be the designata of the sentences? The most natural suggestion is what we might call their *incremental truth conditions*, given the facts that determine the designation of their terms:

> Given that Mark Twain wrote *Tom Sawyer*, (11) is true iff
> Mark Twain grew up in Missouri.
> Given that Mark Twain wrote *Huckleberry Finn* (12) is
> true iff Mark Twain grew up in Missouri.

(11) and (12) agree with respect to the additional requirements they impose on the objects that fit the descriptions in them, the requirements that appear on the right hand side of the biconditionals. These incremental truth conditions can be taken as facts or states of affairs involving the described object - Mark Twain in this case - rather than the descriptive complexes. This proposal we can summarize as follows:

Expression	Designation	
	Value-free reading	Value-loaded reading
Description	Descriptive condition	Object described
Sentence	Truth conditions: State of affairs involving descriptive condition	Incremental truth conditions: State of affairs involving object described

The principles we adumbrated earlier need to be modified:

(B')   Substitution of basic terms that co-designate do
            not affect the truth conditions of a sentence;
            substitution of descriptions does not effect the
            incremental truth conditions of a sentence, but
            may effect the truth conditions.

Substitution': (From (A) and (B')) Substitution of basic co-
            designating terms does not effect what a sentence
            designates.Substitution of co-designating descrip-
            tions does not effect what a sentence designates on
            a value-loaded reading.

(Picking any reasonably coherent notion of truth conditions and sticking
with it will lead to basically the same modification.

We might take the truth conditions of a statement to correspond to
the models in which it comes out true. A model assigns an appropriate
extension to each name and predicate in the language. G3 and G4 clearly do
not pass the test of being true in the same models. They will both be true in
the model that reflects the actual world (given our assumptions). In this
model the object named by a will be a member of the extension of ϕ, so G3
will be true. The pair of objects named by a and b will be a member of the
extension of π, so G4 will be true. But there will be plenty of models in
which one of these assumptions is true but not the other, and so there will be
plenty of models in which G3 is true but not G4. On the other hand, if we
restrict ourselves to the models in which the assumptions $\phi(a)$ and $\pi(a,b)$ are
both true both G3 and G4 will be true.

The substitution principle is also undermined by the propositions of possible worlds semantics. Consider the initial statement in Church's argument. Are we to take Ivanhoe as part of the subject matter or not? If not, we get the set of all possible worlds (or all possible worlds in which Scott exists) as the designation of C1. If we take writing Ivanhoe to be part of the subject matter, we get the set of worlds in which Scott wrote it. These are quite different sets of worlds.)

IV. THE MODIFIED SLINGSHOT

Given these principles we can construct a modified version of the slingshot that purports to show that the lower right hand box of our diagram cannot really be the incremental truth conditions, but must be simply truth values. That is, if we give the descriptions in a sentence their value-loaded reading, we are forced to take the sentence to designate truth values rather than incremental truth conditions.

The modified slingshot purports to show that the incremental or value-loaded designata of all true sentences with descriptions are the same. Let $\iota x\ [\phi(x)]$ and $\iota y\ [\psi(y)]$ be distinct objects and F and G be distinct properties such that it is true that $F(\iota x\ [\phi(x)])$ and $G(\iota y\ [\psi(y)])$. Give all of the sentences in the following sequence a value-loaded reading.

M1.	$F(\iota x\ [\phi(x)])$	Ass.
M2.	$\iota x\ [\phi(x)] = \iota x\ [\phi(x)\ \text{and}\ F(x)]$.	Redist.
M3	$\iota x\ [\phi(x)] = \iota x\ [x \neq \iota y\ [\psi(y)]\ \&\ x = \iota x\ [\phi(x)]]$	Subs.
M4	$\iota y\ [\psi(y)] = \iota y\ [y \neq \iota x\ [\phi(x)]\ \text{and}\ y = \iota y\ [\psi(y)]]$	Redist.
M5	$\iota y\ [\psi(y)] = \iota y\ [\psi(y)\ \text{and}\ G(y)]$.	Subs.
M6	$G(\iota y\ [\psi(y)])$	Redist.

Given our original redistribution principle and the revised substitution principle, M6 should have the same designation as M1, if both are given a value-loaded reading. To block the modified slingshot, we need to turn to the redistribution principle.

V. TRUTH CONDITIONS AND SUBJECT MATTER

In "Semantic Innocence," Barwise and I associated the faults of the redistribution steps with the problem that I have elsewhere called "losing track of subject matter",[9] which affects both the model theoretic and possible worlds conceptions of truth conditions.

All logical truths are logically equivalent; if we take logical equivalence as a criterion for sameness of truth conditions, they will all designate the same thing according to (A). For example (11) and (12)

> (11)   Mary is sitting or Mary is not sitting.
> (12)   Peter is picking peppers or Peter is not picking peppers.

are each true in all models for a language that contains both sentences. So, on the logical equivalence criterion, (11) and (12) have the same truth conditions.

Similarly, if we take necessary truth as our criterion of sameness of truth conditions, (11) and (12) will designate the same thing by principle (A).

Given some fairly plausible assumptions, this means that neither of these conceptions of truth conditions support the notion of truth conditions being *about* a particular object. Consider the following sequence of sentences:

> P1   Peter picked a peck of pickled peppers.
> P2   Peter did not pick a peck of pickled peppers .
> P3   Peter picked a peck of pickled peppers or Peter did not pick a peck of pickled peppers.
> P4   Mary is sitting and (Peter picked a peck of pickled peppers or Peter did not pick a peck of pickled peppers).
> P5   Mary is sitting

Intuitively the truth conditions of P1 are about Peter, since it mentions him and predicates something of him. It seems that if the truth condition of S are about Peter, those of ~S should be too, so P2 is about Peter. If the truth conditions of S and those of Q are both about an object, it seems that those of S *or* Q will be about that object, so those of P3 are about Peter.[10] It seems that if the truth conditions of S are about an object, those of Q *and* S will be about that object, so those of P4 are about Peter. But P4 is logically and necessarily equivalent to P5, so on the conceptions in question, P4 and P5 have the same truth conditions, so those of P5 are about Peter. But this means that the concept of being *about*, on these conceptions, is essentially empty.

The fact that the model theoretic and possible worlds conceptions of truth conditions lose track of subject matter in this way raises problems in many areas, such as semantics of attitude reports.[11] In "Semantic Innocence" Barwise and I diagnosed the slingshot as another place where losing track of subject matter leads to problems. We were mainly concerned with Davidson's versions of the slingshot; he often justifies the redistribution steps by appeal to the logical equivalence of the sentences in question. We also saw the logical equivalence criterion in the background of Church's discussion although he does not appeal to it. Barwise and I criticized logical equivalence as a criterion of sameness of designation, on the basis of considerations like those adumbrated in the last few paragraphs, and on this basis, rejected redistribution steps.

Olson points out that this does not really get to the heart of the issue of redistribution steps.[12] The argument Gödel uses does not rely on the logical equivalence criterion, but on a specific and intuitively plausible claim that two sentences have the same content. The pairs of sentences in the Gödel slingshot that are linked by redistribution steps, such as G2 and G3, do not seem to involve any dramatic changes in subject matter of the sort that the logical equivalence or necessary equivalence criteria permit. So, one can suppose that the criteria of logical or necessary equivalence are too strong for sameness of designation of sentences, while still supposing that redistribution is a correct principle. Indeed, this was presumably Frege's view at the time of the *Begriffschrift*. His carving up content principle would support the principle of redistribution, but he did not claim that all necessary truths or all logical truths had the same content.[13]

VI. TRUTH CONDITIONS AND REDISTRIBUTION

Consider G2 and G3. They are logically and necessarily equivalent, so by these criteria of sameness of truth conditions, they will designate the same object. But this does not seem to turn on losing track of subject matter. G3 brings in new logical apparatus, the definite description operation and the identity sign, but no new terms for objects, properties or relations (other than identity, arguably a part of the logical apparatus).

Olson points out that taking sameness of fact as our criterion does not deliver such a clear answer. Can we really distinguish between the facts described by G2 and those described by G3? The question forces us to recognize two different conceptions of facts, with different criteria of identity. According to what Olson calls the "existential" conception, facts are identical if they necessarily co-exist, while on the "structuralist" conception, they are individuated by sequences of properties and objects.[14] The existential conception supports redistribution, and seems to leave us with no obvious way out of the modified slingshot.

The spirit of the structuralist conception is contrary to principle (C) and to redistribution, and certainly allows one to block the slingshot. Where $a = \iota x\,[\phi(x)]$, a natural structuralist representation of M1 and M2 might be as follows:

<F, a>
<=, a, a>

(Remember that we are dealing only with value-loaded readings). For the structuralist, we have two quite different facts designated, with different structures.

Olson is clearly sympathetic to the existential concept of facts, and to the charge that the structuralist conception improperly mixes metaphysics and syntax. He also thinks the fine-grained structuralist conception of facts will be susceptible to a sort of metaphysical slingshot inspired by Bradley.

VII. THE METAPHYSICAL SLINGSHOT

Olson's conclusion is not quite that that there is only one fact on the structuralist conception, but rather than any fact necessitates every other. Here is what Olson says:

> It seems to me that the structuralist approach is exposed
> to objections of the sort that Bradley made against
> Russell. For it holds that an object a has the property F if
> and only if there exists a fact whose constituents are a and
> F. Bradley's question was what manner of thing a could
> possibly be. Is it an ordinary thing, replete with all its

properties and relations to other things? If so, how can its having F consist in its being the constituent of a fact of which F is another constituent? Its identity as a constituent depends on its properties, including F. Moreover, a is supposed to be the same in each of the facts of which it is a constituent. If the fact that a has F is determined by its constituents, and a, which is one of them, has the property G, how can this fact be compatible with a's not being G? It is true that distinct facts can necessarily coexist according to the structuralist. But since G can be any property of a, he now appears to be committed to what James called "the Block Universe." Each fact that exists is incompatible with any state of affairs that does not exist. For the state of affairs of a's not being G would have as a constituent a, an object which is G. The structuralist could meet this objection by taking a as a "bare particular," but I do not think that Bradley would be alone in regarding this as a form of theoretical suicide. [15]

This is pretty succinct. Let's see if we can spell it out a little. We'll just consider simple facts, involving an object having a property or a sequence of objects standing in a relation. I'll assume that there are both positive and negative facts; for example, there is the fact that I am sitting, and the fact that I am not standing. And I'll use *states of affairs* for fact- like things that are not facts. So my sitting and my not sitting are both states of affairs - I'll call them opposites - and one is a fact. A very basic metaphysical principle is that of two opposite states of affairs, one and only one is a fact.

Suppose that f is the fact that I am sitting, and f' is the fact that I am from Nebraska. Then, since I, as a member of f, am replete with all my properties, the first fact somehow contains the second. According to what I shall call Olson's principle, f is incompatible with the opposite of f'. Then, by the basic metaphysical principle mentioned above, f necessitates f'. And by parity of reasoning, f necessitates every other fact with me as a constituent. If we suppose that every object stands in some relation or other to every other object, and that every fact has some object as a constituent, we get the result that f necessitates every fact. So, given Olson's principle, this conception of facts seems to lead to something quite aptly called a "Block universe".

What about Olson's principle? To make it plausible, I think we need to make two assumptions about Olson's intentions. First, we need to assume that G is supposed to be an essential property of me. Only the facts essential to my existence are necessitated by facts with me as a constituent. Arguably, I would not be me, unless I had the very parents that I had. So any fact with me as a constituent, necessitates the fact that I have the parents that I do. But then, it seems, all sorts of facts about parentage are going to be necessitated, about the parents of my parents, and their parents, and so forth. We seem to have not so much a block universe as a sort of lace universe, with the fact that I am sitting necessitating an odd assemblage of facts back to facts about Adam and Eve and Cain or Abel and the like, but with lots of unnecessitated holes of random size and shape in between. But the lace universe is enough of a problem for the structuralist theory of facts.

The second assumption we need to make about Olson's intentions is that he is assuming that facts are the basic building blocks of one's ontology. On this idea, objects must derive their existence and identity from the facts in which they are involved. To then individuate facts in terms of objects, as the structuralist proposes to do, seems to be ruled out. If one is thinking of the objects "replete with their properties," prior to facts, then facts are not basic. If one is thinking of the object as bare particulars, then one surely is committing theoretical suicide, since now it is these bare particulars that are ontologically basic.

VIII. FACTS AND SITUATIONS

I want now to describe a view that rejects this second assumption. This is the view in *Situations and Attitudes*, although we chose in that work not to reflect the whole view in the formal apparatus. More recent versions of situation theory incorporate the entire view into the formal apparatus.[16] This view takes something like Olson's existentialist view about situations, but his structuralist view about facts. I then consider a problem for this view related to the metaphysical slingshot, and suggest a solution to the problem.

First let's adopt a little notation and terminology. By a scheme of individuation and classification I mean a domain of individuals together with a domain of relations. Where R is an n-ary relation and $a1,...,an$ is an appropriate sequence of objects, I shall call $R,a1,...,an$ an issue and an issue together with 1 or 0 a state of affairs. A scheme of individuation determines a set of issues.

$$R,a_1,...,a_n;1$$

is the state of affairs that is a fact if and only if a1,...,an stand in the relation R, while

$$R,a_1,...,a_n;0$$

is its opposite, the state of affairs that is a fact if and only if a1,...,an do not stand in the relation R. A basic metaphysical principle is that a state of affairs is a fact if and only if its opposite is not one. But what determines which states of affairs are facts? One answer might be that this is bedrock. Some just are, some are not. This is not my view, however. It seems to me that we have a notion of a reality or realities that might be individuated and classified in different ways, according to different schemes of individuation. Imagine a checkerboard, for example. We could individuate it as sixty-four squares, and classify them with the properties of being red and black, and the relations of directly under and directly to the right of. Or we could individuate it as eight rows and eight columns, and classify them in terms of the relations of being red at and being black at, being under and being to the right of. There are two different ways of getting at the same reality. I want to say that the checkerboard can be considered as a situation, which, given one scheme of individuation and classification, determines some states of affairs to be facts, and given another, determines other states of affairs and facts. Reality transcends any one scheme of individuation.

On this conception, then, facts are not basic entities from which the world is constructed, but more like the accurate measures of situations, relative to a standard of measurement. Being a fact is really a derivative property. Instead of the fundamental property of being a fact, we have the fundamental relation of being determined to be a fact by a situation.

$$s \models \sigma$$

is read "situation s makes state of affairs σ factual.

Where $\sim\sigma$ is σ's opposite, two plausible principles are:

a)   If $s \models \sigma$, then not: $s \models \sim\sigma$

b)   If there is an s such that $s \models \sigma$, then there is no s' such that $s' \models \sim\sigma$.

a) says that no situation makes a state of affairs and its opposite factual; b) says if one situation makes a state of affairs a fact, no other situation can make its opposite a fact.

A third plausible principle (persistence) is:

c)   If s is part of s' and s $\models \sigma$, then s' $\models \sigma$.

The principle that there is a world we could express this way

d)   There is an s such that for all σ, s $\models \sigma$ or s $\models \sim\sigma$

The world is not the totality of facts, but something from which the objects and relations that are used to construct states of affairs - both the facts and the non-facts - are abstracted. The additional situations I have in mind are not alternative realities to the actual world, but parts of it. We could speak, for example, of the situation in this room during the present hour.

IX. INDIVIDUATING SITUATIONS

Looking at things this way, it seems unproblematic to individuate facts in terms of their constituents. But what of situations? How do we individuate them? Unless we answer this question, can we even give an answer to the question of whether there is really more than one?

In our book *Situations and Attitudes*, Jon Barwise and I used the term "situation" both for what I am calling situations and for what I am calling states of affairs. We called the former "real situations" and the latter "abstract situations". Among abstract situations, we distinguished between factual and non-factual. So factual abstract situations were what I am calling facts or sets of them.

One can distinguish between the internal and the external properties of situations. The internal properties are which states of affairs they determine to be factual. An example of an external property is being perceived by a person at a certain time. Another is leaving a certain issue open, being "undefined". Given principle a) above, one would know that s is not identical to s' if s determines σ to be a fact and s' determines $\sim\sigma$ to be a fact. But given principle b), this test will never apply.

However, if one thinks that situations have not just fact-determining role, but a fact- constitutive role, there will be lots of ways of individuating them. It is natural to take situations to be the constituents of facts involving perception, causation, and the like.

I want to briefly explore a certain problem related to these two roles for situations, related to the issues Olson raised. Basically the problem is that situations that are big enough to determine states of affairs to be facts are too big for most of the fact-constitutive uses for situations. Consider the fact that I am sitting (as I write this). How big does the situation have to be, that determines this to be a fact? At first glance, one would think that the situation in the room, at the present time would do it. That situation includes everything that is going on in the room at that time. Is that not enough to settle the issue of whether I am sitting or not?

At second glance, it does not really seem big enough. Let's go back to the checkerboard for an analogy. Consider the fact that row 2 is red at column 2. At first glance, it seems that just the situation at square 10 is what settles this issue. But at second glance, this does not seem right. That little patch could have been red, even if there were no rows or columns, or not the ones that there are. That patch could have been part of a board that was not suited for checkers at all, in which geometrical shapes of various colors shapes and sizes lie adjacent to it in a random way. Given the rest of the board, which guarantees the existence of the row and column and question, what goes on at that square is what settles the issue of the color of row 2 at column 2. But what goes on at square 10 does not settle it all by itself, because what goes on at square 10 does not suffice by itself to establish the existence of the column and row that are constituents of the fact.

In a way, that fact that square 10 is red, and the fact that row 2 is red at column 2, seem like the same fact. But they are not, since they have different constituents. And it is only the former fact, not the latter, that is really settled by what goes on at square 10.

The situation in the room seems similar. Just as the column and row "transcend" the square, and are not established to exist just by what goes on in the square, so I transcend the situation in my study. In some sense, things could be just as they are here, and I not exist. We do not seem to be able to find that possibility, so long as we "measure" the situation with the scheme of individuation with persons in it. But if we think in terms of "person-stages", or in terms of rooms and their properties (there is a person of such and such a type writing in this study at this time) we can find them.

It seems that what is going on in this study now, the situation in this study, could be fit into various larger situations.

How large will the situation have to be, then, that settles the fact that I am now writing? It seems it will have to stretch back far enough to establish my existence. But perhaps it will have to stretch even further than that. The problem of the lace universe reasserts itself at the level of situations.

The solution to the problem, I believe, is to grant it. The situations which establish the ordinary facts we are interested in may have to be quite large. This does not show that there are no smaller situations, however, just that the smaller ones, the ones we perceive and the like, do not establish quite as much as we thought they did.

The right way to look at it, I think, as roughly as follows. The small situations we perceive, examine, and the like do not establish the facts we are interested in by themselves. They establish them only relative to other facts. The key relation is not "determines to be factual," but "determines to be factual given certain facts". This has to do with the incrementality of situations. I want to know what the situation in this room supports about a certain filing cabinet, a certain typewriter, a certain person. I identify these individuals not by being privy to all of the fact that determine their identity, but by being in a relationship with them.

So, I suspect, the notion we need is that of *incrementally determines to be a fact*. That is, $s \models \sigma$, given σ'. I will end on this somewhat indecisive and incomplete note.

NOTES

[1]   The research reported in this paper has been made possible by a gift from the System Development Foundation to the Center for the Study of Language and Information at Stanford. The paper was read at the 1990 Pacific Division Meetings of the American Philosophical Association in a symposium with Ken Olson and John Boler on Olson's book, *An Essay on Facts* (Olson, 1987). I learned a great deal from Olson and Boler's presentations, and from a point made by Mark Richard. I have also had many other rewarding conversations with Olson on the topics of this paper. The paper corresponds to a portion of the workshop on situation semantics that I gave with John Etchemendy at the San Sebastian Conference. Etchemendy and Elizabeth Macken gave me helpful comments on the penultimate draft.

[2]   One should really talk about uses of sentences, since (most) sentences do not have either truth conditions or truth values --- the two candidates for

designata we consider --- except as used on a specific occasion. But since I won't discuss issues of context sensitivity, I'll stick with "sentence," which fits with most of the literature on the slingshot.

[3]   Olson discusses Frege and the slingshot, Olson, 1987, pp. 65-82.

[4]   Church 1956, p. 25.

[5]   Quine 1976, pp. 163-64.

[6]   Davidson 1967.

[7]   Barwise and Perry 1981.

[8]   Olson 1987.

[9]   In Perry, 1989.

[10]   As Olson points out (Olson, 1987, p 85), Wittgenstein's doctrines in his *Tractatus* do not agree with this, since he didn't think that tautologies had any content at all. See also Partee, 1989.

[11]   See Partee, 1989 and Perry 1989.

[12]   Olson, 1987, pp85 ff.

[13]   See Weitzman, 1989 for a discussion of Frege's "carving up content" principle.

[14]   Olson, 1987, pp91ff; Olson's distinction and terminology is derived from Kit Fine's distinction between structuralist and empiricist conceptions of facts in Fine 1982.

[15]   Olson, 1987, pp. 99-100.

[16]   See Barwise,1989 .

BIBLIOGRAPHY

Barwise, Jon.: 1989, *The Situation In Logic*. Stanford: CSLI/University of Chicago Press.

Barwise, J. and Perry, J.: 1981, Semantic Innocence and Uncompromising Situations. *Midwest Studies in the Philosophy of Language*. VI.

Church, Alonzo: 1956, *An Introduction to Mathematical Logic*. Princeton: Princeton University Press.

Davidson, Donald: 1967, The Logical Form of Action Sentences. In Nicholas Rescher, ed., *The Logic of Decision and Action*. Pittsburgh: University of Pittsburgh Press.

Fine, Kit: 1982, First-Order Modal Theories III --- Facts. *Synthese* **53**.

Gödel, Kurt: 1972, Russell's Mathematical Logic. In David Pears, ed., *Bertrand Russell: A Collection of Critical Essays*. Garden City, N.Y.: Anchor Books.

Olson, Ken: 1987, *An Essay on Facts*. Stanford: CSLI/University of Chicago Press.

Partee, Barbara: 1989, Speaker's Reply. In Sture Allen, ed., *Possible Worlds in Humanities, Arts and Sciences*. *Proceedings of Nobel Symposium 65*. Berlin/New York: Walter de Gruyter.

Perry, John: 1989, Possible Worlds and Subject Matter. In Sture Allen, ed., *Possible Worlds in Humanities, Arts and Sciences*. *Proceedings of Nobel Symposium 65*. Berlin/New York: Walter de Gruyter.

Perry, John: Forthcoming. *The Problem of the Essential Indexical and Other Essays*. New York: Oxford University Press.

Quine, W.V.: 1976, Three Grades of Modal Involvement. In W.V. Quine, *Ways of Paradox*, revised and enlarged edition. Cambridge: Harvard University Press.

Weitzman, Leora: 1989, *Propositional Identity and Structure in Frege*. Doctoral Dissertation, Stanford Philosophy Department.

Philosophy Department
Stanford University

ALFONSO GARCIA SUAREZ

REFERENCE WITHOUT SENSE: AN EXAMINATION OF PUTNAM'S SEMANTIC THEORY[*]

I. ARE MEANINGS IN THE HEAD?

In several writings[1] Hilary Putnam has stated his rejection of traditional semantic theories by means of the slogan "Meanings ain't in the head". The slogan is deceptive. Out of context, it could serve as a lemma for semanticians with very different backgrounds. Thus, a Fregean philosopher could endorse it as a statement of his antipsychologistic stance. Although Putnam acknowledges that Frege and Carnap identified meanings with abstract entities, not with mental ones, he thinks that the clash between psychologism and Platonism is "a tempest in a teapot, as far as meaning-theory is concerned",[2] since understanding the meaning of a word, grasping the abstract entity which is its sense, would still be an individual psychological state. And so Frege would share with traditional semantics an assumption that Putnam is going to reject:

> (I)   That knowing the meaning of a term is just a matter of being in a certain psychological state.

Similarly, Putnam's slogan would trigger a warm reaction in a Wittgensteinian philosopher. But appearances are misleading again, since the Wittgensteinian too could allow assumption (I) without more ado if "psychological state" is read in a rather liberal sense. A sense wide enough to content such a philosopher is the one that Putnam gives to this expression when he counts *dispositions* among such states[3] or when he asserts that the knowledge involved is "knowledge in the wide sense, implicit as well as explicit, 'knowing how' as well as 'knowing that', skills and abilities as well as facts".[4]

115

A. Clark et al. (eds.), Philosophy and Cognitive Science, 115–133.
© 1996 *Kluwer Academic Publishers.*

Assumption (I) is presented as the first of "two unchallenged assumptions" on which, according to Putnam, traditional semantic theory rests. The second assumption is:

> (II)   That the meaning of a term (in the sense of 'intension') determines its extension.

Putnam contends that these two assumptions are not jointly satisfiable by any notion, and hence that the traditional concept of meaning is radically defective.

In order to sustain these contentions he argues as follows: Assumption (I), *adequately interpreted*,[5] leads to the consequence

> (C1)   That the speaker's psychological state determines a term's intension.

And (C1) and (II) lead to the consequence

> (C2)   That the speaker's psychological state determines a term's extension.

Then Putnam offers two kinds of thought-experiment designed to show that (C2) is false. According to him, a term's extension is not fixed by the speaker's mental state for two reasons: (a) because extension is *socially* determined - there is a division of linguistic labour -; and because it is *indexically* determined - the environment itself contributes to determining the reference of a speaker's or a community's words. This leaves two open courses: either giving up assumption (I) or giving up assumption (II). In "The Meaning of 'Meaning'", Putnam endorses assumption (II) and chooses the first alternative. Thus, the actual clash between traditional semantics and Putnam's new theory of reference concerns assumption (I).

In this paper I will try to show, firstly, that the hypothesis of the division of linguistic labour is substantially right but constitutes no serious challenge to a traditional theory of meaning, and, secondly, that the indexicality thesis is not correct; nevertheless, if we admit it, it leads us to an option: either assumption (II) cannot be maintained, or, if assumption (II) is held immune, then Putnam's theory becomes a variety of traditional semantics.

II. THE DIVISION OF LINGUISTIC LABOUR.

As Putnam puts it, assumption (I) is somewhat ambiguous. Under a first reading, (I) would assert that being in a certain psychological state is a *necessary* condition for knowing the meaning of a term - the weak interpretation. Now, several passages in Putnam's writings heavily support the guess that he has in mind a stronger interpretation. When read in this strong sense, assumption (I) runs so:

(I_s) That being in a certain psychological state is a
            sufficient condition for knowing the meaning
            of a term.

Under this interpretation, Putnam's argument referred to in #1 becomes the following one: Assumption (I_s), *adequately construed*,[6] entails (C1); and, by assumption (II) it entails also (C2); but (C2) is false, and, as (II) is to be maintained, we must conclude, by *modus tollens*, that assumption (I_s) is false.

If (C2) holds, i.e. if the mental state of the speaker determines the reference of the terms that he uses, it follows that it is impossible for two speakers to be in the *same* mental state and that the reference of the same term A in the idiolect of the first one is *different* from its reference in the idiolect of the second one. Of the several counterexamples that Putnam presents against this alleged impossibility, some involve the division of linguistics labour. So let us suppose that I cannot tell the difference between an elm and a beech. We still say that 'elm' has in my idiolect the same reference it has in anyone else's - the set of elm trees - and that the set of beeches is the reference of 'beech' both in my idiolect and in that of all English-speakers. But, since my concept of an elm is *exactly the same* as my concept of a beech, the difference between the references or extensions of 'elm' and 'beech' in my idiolect cannot be the result of a difference of concepts. What this would show is that the determination of the reference is in this case social and not individual: "What I refer to as an 'elm' is, with my consent and that of my linguistic community, what people who can distinguish elms from other trees refer to as an elm".[7] The average speaker's linguistic competence depends in such a case on the *expert*, a special subclass of speakers. Consequently, when a term like 'elm', 'beech', 'aluminium', 'gold', etc. is subjected to the division of linguistic labour, what determines its extension is not the average speaker's psychological state: "it is only the sociolinguistic state of the collective linguistic body to which the speaker belongs that fixes the extension".[8]

The conclusion that Putnam draws is that reference is underdetermined by the individual speaker's concepts. I will accept that the hypothesis of the division of linguistic labour is substantially correct and that it forces us to reject the corollary (C2). Now, does it also force us, as Putnam contends, to reject by *modus tollens* assumption (I_s)?

Putnam states his rejection of (I_s) in different ways. So we read in one passage that "the extension of a term is not fixed by a concept that the individual speaker has in his head", and few lines later, that the meaning "cannot be identified with 'intension' (...) if intension is something like an individual speaker's *concept*".[9] Although Putnam here stresses the word 'concept', I would like to emphasize the expressions 'that the *individual* speaker has in his mind' and '*individual* speaker's'. For it is somewhat revealing that, after establishing the indexicality thesis, these qualifications do not appear. Thus at the end of the section on "Indexicality and Rigidity" he tells us that the theory that natural-kind words like 'water' are indexical leads to the consequence - if we accept assumption (II) - that a "difference in extension is *ipso facto* a difference in meaning for natural-kind words, thereby giving up the doctrine that meanings are concepts, or, indeed, mental entities of *any* kind".[10] And the question is: Does the hypothesis of the division of linguistic labour discredit only the view that meanings are identical with, or are fixed by, concepts *of the individual speakers* or does it also discredit the wider view that meanings are determined by concepts *tout court*, whether these are individual or social property?

III. THE ASSUMPTION OF METHODOLOGICAL SOLIPSISM:
ANOTHER VICTIM FOR THE *MODUS TOLLENS*.

I will try to show that the more general view referred to above is not seriously threatened by the hypothesis of the social determination of reference. With this aim, I would like to point out that in the argument by Putnam outlined at the beginning of #2 the first premise is somewhat elliptical. We have already emphasized twice that the corollary according to which the speaker's psychological state determines extension is entailed by assumption (I_s) *only if* this assumption is *adequately construed*. The interpretation that Putnam has in mind is the interpretation according to which 'psychological state' is to be taken not in the wide sense in which "a psychological state is simply a state which is studied or described by psychology"[11] but in a narrow sense. The psychological states referred to in the premises are only those allowed by what Putnam calls the assumption of *methodological solipsism* (MS), viz. "the assumption that no psychological state, properly so called, presupposes the existence of any individual other than the subject to whom the state is ascribed".[12] So, only when assumption (I_s) is restrictively construed according to MS, the refutable corollary follows from it. But this amounts to saying that (C2) is entailed only by the conjuction of (I_s) and (MS). In fact, although this is an obscure point in Putnam's writings, in "The Meaning of 'Meaning'" the derivation of (C2) goes as follows:

> Let A and B be any two terms which differ in extension. By assumption (II) they must differ in meaning (in the sense of 'intension'). By assumption (I), *knowing the meaning of A* and *knowing the meaning of B* are psychological states *in the narrow sense* - FOR THIS IS HOW WE SHALL CONSTRUE ASSUMPTION (I). *But these psychological states must determine the extensions of the terms A and B just as much as the meanings ('intension') do.*[13]

But now let us suppose that we do not construe assumption (I) like this. Suppose that (I_s) is held but MS is rejected. To come back to the 'elm' - 'beech' case, when I use the word 'elm', although I cannot tell an elm from a beech, I still *know* that elms and beeches are different kinds of tree and I *know* that other speakers - the experts - possess differentiating criteria or abilities. But then - I would like to contend - my *concept* of an elm is *not* exactly the same as my concept of beech. My knowledge that the extension of 'elm' differs from the extension of 'beech' is *conceptual* knowledge. My concept of an elm is, let us suppose, that of a big deciduous tree which is not, *inter alia*, a beech and which experts can distinguish from beeches. Now, this last qualification implies that my concept of an elm cannot be identified with a psychological state in the narrow sense, for it makes reference to other individuals besides myself - the experts.

In his recent book *Representation and Reality*,[14] Putnam allows that I may know that elms and beeches are different species of tree. Then, he says, my mental representation of an elm would include that it is not a beech, and *mutatis mutandis* the same holds with respect to my mental representation of a beech. But, he argues, "what this *amounts to* is that my mental representation of an elm includes the fact that there *are* characteristics which distinguish it from a beech",[15] and similarly with respect to beeches. But since *ex hypothesi* my respective mental representations do not include an specification of these characteristics, my mental representation of an elm trivially differs from my mental representation of a beech only in the phonetic shape of the names, in my knowing that the first species is *called* 'elm' and the later species is *called* 'beech'.

Assumption (I) is restated in *Representation and Reality* in Aristotelian-Lockean form as follows: "Every word he uses is associated in the mind of the speaker with a certain mental representation". Now, putting the issue in terms of mental representations does contribute to darken it because it suggests pictures or images in individual brain/minds and so it does confine traditional semantics to an indefensible mentalistic version. It has indeed the same effect as the assumption of methodological solipsism had in "The Meaning of 'Meaning'". But, as Putnam himself acknowledges in the case of stereotypes,[16] conceptual contents do not need to be pieces of imagery: they could be beliefs stated in words. Now, concerning elms, I believe that they are a species of big deciduous tree, that they grow in Europe,...and that they are not beeches. Given this last belief, I would be prepared to assent to sentences like 'If all the trees in the garden are elms, then there are no beeches in it', and to dissent from sentences like 'If the garden contains only elms, then it necessarily contains beeches'. These and other similar dispositions do show that my concept of an elm and my concept of a beech are different. Certainly, although my respective concepts do include knowledge *that* there are distinguishing characteristics, they do not include knowledge of *what* these characteristics are. But Putnam's complaint about this rests on the false assumption that two concepts can differ only if they include an explicit specification of the criteria of distinction for the objects that fall under them. This assumption might be correct if concepts were to be identified with mental representations in the narrow sense, but it is not valid for concepts in the liberal sense referred to above.

No doubt, one of the reasons why I know that elms are not beeches and *vice versa*, is that I know that the experts can distinguish them because they know the distinguishing characteristics and I rely on their expertise.

Putnam is on the right track when he stresses the division of linguistic labour. Now, this means that my concept of an elm includes the content of the description

(D)   species of tree which the experts can distinguish from beeches.

Putnam apparently contends that (D) would amount to

(D')   species of tree which is *called* 'elm' by English-speaking experts.

But, he objects, a metalinguistic description as (D') cannot give the meaning of 'elm' because it is not a *translation* of that word - that elms are called 'elms' is not part of the concept of an elm. But if (D') is not synonymous with 'elm', it cannot be identified with its meaning. Once again the conclusion would be "the impossibility of identifying meanings with the descriptions that the speakers 'have in their heads', i.e. of identifying the notions of *meaning* and *mental representation*".[17]

Now, in the first place, there is no reason why (D) should be *synonymous* with 'elm'. It would be enough that it were one ingredient of a cluster of descriptions that collectively fix the reference of the word. That cluster could include, *inter alia*, the stereotype which is associated with the term. In the second place, (D) does certainly builds the division of linguistic labour into my concept, but there is no need to reduce it to (D'). First of all, the experts on which I rely do not have to be speakers of my language. They could be French-speaking persons. Putnam would counter that (D) would be than reducible to

(D'')   species of tree which is *called* 'orme' by the French-speaking experts,

that has the same shortcomings as (D'). But it well might happen that, although I know that there are experts which can tell elms from beeches, I do not know what language these experts speak. In such a case Putnam could not try to reduce (D) to any metalinguistic form because I would not know what the word is that the experts use. What this show is that Putnam's reduction of (D) to some metalinguistic version is wrong. Moreover, the experts on which I rely do not have to be speakers of *any* language. Perhaps it could rely on the discriminatory abilities of some creatures that do not have developed any language.

Putnam thinks that the view that we have defended is an "heroic" way out that can be refuted by constructing a Twin Earth example.[18] He asks us to suppose that the words 'elm' and 'beech' are *switched* in Twin Earth - a planet exactly like Earth save for the circumstance that in it 'elm' is the name for beech and 'beech' is the name for elm. Suppose now that I have a *Doppelgänger* in Twin Earth who is type-identical with me and such that his psychological states are identical to mine. However, the extension of 'elm' is *elm* in my idiolect but *beech* in my *Doppelgänger*'s idiolect. He means *beech* when he says 'elm' and I mean *elm* when I say 'elm'.

But now it happens that the modified Twin Earth example is again vulnerable to the objection that my psychological states and my *Doppelgänger*'s psychological states are *not* exactly type-identical. The contents of our respective psychological states, of our propositional attitudes, are not identical when I say 'I believe that there are elms in the garden nearby' and when my *Doppelgänger* says the same words. In fact, when I say 'elms' I mean elms and when my *Doppelgänger* says 'elms' he means beeches, although we both utter the same word. But this fact affects the oblique occurrences, within 'that'-clauses, which make up the contents of our respective psychological states. I believe that there are *elms* in the garden nearby; my *Doppelgänger* believes that there are *beeches* in the garden nearby. Tyler Burge[19] has drawn attention to this fact, pointing out that psychological contents are affected by the nature of the social and physical environment, although the physical and qualitatively mental features of the individual are still identical. That is, if Burge's argumentation is correct, it makes no sense to suppose that my *Doppelgänger* and I are in these circumstances, exact duplicates as far as our thoughts, beliefs, etc.. are concerned.

But, independently from the question whether my concept of an elm is the same as my concept of a beech and whether my psychological states and those of my *Doppelgänger* are identical, the supporter of traditional semantic theory can still argue that, although Putnam's thought-experiment may establish that meanings are not fixed by concepts of the *individual* speaker, it has no strength against the modified view according to which meanings are determined by the *conceptual pool* of the linguistic community at large - including the experts.

We saw earlier that Putnam points out two courses open if and once we are convinced that extension is not fixed by the individual speaker's concepts. But now it happens that there is another different course open that Putnam fails to notice. We can either give up assumption (I_s), as Putnam suggests, or give up only the assumption of methodological solipsism. And since "three centuries of failure of mentalistic psychology is tremendous evidence"[20] against methodological solipsism, let us abandon it without regret!

Once purged of the assumption of methodological solipsism, a traditional semantic theory - for instance, a sense theory of Fregean flavour - is not seriously threatened by the hypothesis of the division of linguistic labour. Of course, the theory that Frege historically put forward leaves no place for social cooperation. As Michael Dummett has accurately noted, Frege thinks of language as an overlap of idiolects. But, once we admit that a word like 'elm' does not get its meaning from concepts of individual ownership but from the communal knowledge belonging to the linguistic community as a whole, a sense theory of Fregean style can easily accommodate the phenomenon of the division of linguistics labour.

IV. THE CONTRIBUTION OF THE ENVIRONMENT

If traditional semantics does after all accommodate the division of linguistic labour, the Fregean thesis that reference is determined by sense - now conceived of as a cognitive but socially-determined notion - in no longer in jeopardy. For Putnam's argumentation against such a theory to be effective it should have to establish that extension is also underdetermined by the mental states of *all* language users - including the experts. And it is just this what Putnam claims to be established by the thesis of the indexical determination of the reference of natural-kind words. The indexicality of these words would show that their reference is not fixed even by the conceptual pool of the linguistic community, that meaning is not a conceptual matter at all, and that it is necessary to appeal to some kind of external relation.

Putnam puts forward a counterexample to the thesis that it is impossible for two linguistic communities to have the same pool of intensions or concepts if their terms have different extensions. Suppose that Twin Earth is again exactly like Earth; there are even English-speakers there. But there is a difference: in Twin Earth the liquid which fills lakes and rivers and Twin-Earthians call 'water', although superficially indistinguishable from water, is not really H_2O but a liquid whose strange chemical formula we shall represent as XYZ. There is an important difference between this example and that of the 'elm'-'beech' pair above. Earthian botanists might easily discover that 'elm' in the Twin-Earthian dialect has the set of beeches as its extension. But now let us come back to 1750, when Daltonian chemistry had not yet appeared either in Earth or in Twin Earth: *nobody* in Earth nor in Twin Earth could at this time distinguish the liquid in one planet from the liquid in the other. Now the confusion involves the whole linguistic community. Let Oscar$_1$ and Oscar$_2$ be, respectively, a typical English-speaking Earthian and a typical English-speaking Twin-Earthian. Then, Putnam contends, it is plausible to suppose that Oscar$_1$ and Oscar$_2$ share the same psychological states and, nevertheless, the extension of 'water' in the mouth of Oscar$_1$ is the set of all wholes consisting of H_2O molecules, whereas the extension of 'water' in the mouth of Oscar$_2$ is the set of all wholes consisting of XYZ molecules.

What this example would establish, according to Putnam, is that the extension of 'water' is not fixed even by the conceptual pool which the linguistic community possesses as a group. The extension of 'water', like that of other natural-kind terms, is determined *indexically*:

> Water is stuff that has the same microstructure as most of the paradigm water; and paradigm water is *paradigm-for-us*, is water in *our* environment (...) What 'water' refers to depends on *the actual nature of the paradigms*, not just on what is in our heads.[21]

V. IS WATER H_2O IN ALL POSSIBLE WORLDS?

The doctrine that a word like 'water' contains an indexical component roughly amounts to postulating that we can define 'water' like this: *Whatever bears the relation same-liquid-as (same$_L$) to THIS* (indexically demonstrated) *substance*. The indexical component would enter trough contextually identified paradigms. However, the relation of *sameness$_L$*, or, more generally, *sameness of nature*, is indeed extremely vague. We are told that two samples of liquid are en the relation same$_L$ if they "agree in important properties".[22] But the notion of importance is again hopelessly vague - Putnam himself acknowledges that it is an *interest-relative* notion. So let us introduce interests in Putnam's *Gedankenexperiment*.

Suppose that although in 1750 Daltonian chemistry had not yet been developed, there was still a long tradition of contacts by space travels between the inhabitants of Earth and those of Twin Earth. (If the reader finds such a development of space aircraft without a development of chemistry *physically* impossible, he must recall that it is enough for our purposes if the example offers a mere *logical* possibility, or he can imagine that spaceships were supplied by people coming from a more advanced civilisation.) When chemistry is developed it is discovered that the thirst-quenching, colourless, tasteless substance that fills lakes and rivers in Twin Earth - and that when exported in bottles is very popular among drinkers in Earth - is not H_2O but XYZ. Now, given the *importance* that water has in the lives of the inhabitants both of Earth and of Twin Earth - the reader can imagine at this point that water is part of some quasi-religious ritual of brotherhood between the two communities -, why should we conclude that the liquid in Twin Earth is *not really* water? Could we not say instead that *there are two kinds of water*, each one with a different chemical composition? Why should criteria of chemical composition have an absolute priority over other criteria - of function, role in the ritual, etc.? What semantic sin would be committed by the hypothetical Royal Academy of the Unified language of Earth and Twin Earth or by the hypothetical *Oxford and Twin Oxford English Dictionary* if they were to define: *"Water.* n. Transparent, odourless, tasteless liquid which forms rain and runs in streams and rivers. It also forms with certain proportion of salt the content of seas. It is a compound of hydrogen and oxygen in its Earthian variety and of XYZ in its Twin Earthian variety." Just as in the real world there are two varieties of jade - jadeite and nephrite - and even two varieties of water - "normal" water and heavy water -, so in this possible world there would be two varieties of water and one of them would not be H_2O. If this is really a possible state of affairs, then Putnam's thesis is not right: contrary to what he claims, there are possible worlds in which *water* is *not* H_2O.

In *Reason, Truth and History*,[23] Putnam modifies his example trying to avoid this kind of argument. The liquid in Twin Earth needs not be *that* similar to water. Suppose it is a mixture of 20% alcohol and 80% water, but that the body chemistry of Twin-Earthians is such that they do not feel the difference. Then Twin-Earth 'water' will not exhibit the same observable behaviour as normal water; yet a typical Twin-Earthian speaker might be unaware of these differences and be in exactly the same psychological state as a typical Earthian speaker in 1750.

I think, firstly, that once again Burge's point is a effective against the modified example. The mental states of Oscar$_1$ - a typical Earthian speaker in 1750 - and his *Doppelgänger* in Twin Earth, Oscar$_2$, would *not* be exactly type-identical, since the differences between what Oscar$_1$ and Oscar$_2$ mean by 'water' affect the oblique occurrences, in 'that'-clauses, that supply the contents of their respective propositional attitudes. Secondly, Putnam's reply presupposes that there are two conditions for being real water: having the same ultimate constitution and exhibiting the same behaviour. However, normal water is not a pure substance - it contains salt and other components. Thus, why not count that Twin-Earthian mixture as water? On the other hand, there is no clear-cut answer to the question: What constitutes *sameness of behaviour*?

But now suppose, for the sake of the argument, that Putnam is right and that the correct description of the examples requires us to say that the "water" in Twin-Earth is not really water but a sort of "fools' water". In such a case, Putnam would find himself in a dilemma: either, contrary to his contention, assumption (II), that intension determines extension, would not hold, or, if this assumption is held, then Putnam's theory decays into a variant of traditional semantics.

VI. IS THE INDEXICALITY THESIS COMPATIBLE WITH ASSUMPTION (II)?

Let us take the first horn of the dilemma. If the reference of the term 'water' is determined neither by an individual speaker's concept nor by the conceptual pool of the linguistic community, but "*the natural kinds themselves* play a role in determining the extensions of the terms that refer to them"[24] how can then Putnam maintain assumption (II), according to which intension determines extension? The way in which Putnam severs this conceptual knot is indeed a Pickwickian one. Thus in "Is Semantics Possible?" we are told that

> Meaning indeed determines extension; but only because extension (...) is, in some cases, "part of the meaning".[25]

And in "The Meaning of 'Meaning'" it is claimed that, instead of saying that 'water' has the same meaning in Earth as in Twin Earth, but different extension, and giving up (II),

> it seems preferable to take a different route and identify 'meaning' with an ordered pair (or possibly an ordered *n-tuple*) of entities, *one of which is the extension* (...) Doing this makes it trivially true that *meaning determines extension.*[26]

Of course, if the meaning of an expression is identified with a vector one of whose components is the extension, then the thesis that meaning determines extension is automatically and trivially verified - in fact, it is made true by *fiat*. But now the question is: Is *this* a *theory* that *intension* determines extension? What evidence is there that meaning - in the sense of Putnam's meaning-vector - *is* one of the senses of the word 'meaning'? 'Meaning' never means meaning-vector. Unless Putnam wants his ordered n-tuple, one of whose components is extension, to be identified with the intension of a term, he has not saved the assumption that intension determines extension: he has only decreed that the set-theoretical entity one of whose components is extension determines extension.

The point is that the notion of intension - or sense - is an epistemic one. The intension of an expression has to do with what the speakers know when they understand it - with its cognitive value. This is how Frege introduced the notion of sense in order to account for the existence of true but informative identity statements. Now, in Putnam's semantic theory, what play this role are stereotypes. But stereotypes do not fix reference, according to the indexicality thesis. Putnam's Pickwickian way out, to reconcile the indexicality thesis with assumption (II), is replacing the notion of meaning with a set-theoretic construct. But, literally taken, the thesis that the meaning of a natural-kind word is an ordered n-tuple one of whose components is its extension is simply false. Are we to accept that Lake Michigan "with all its waters" is a component of the meaning of the word 'water'? Obviously it seems more plausible to assume that reference - extension - is *not* an ingredient of meaning.[27]

The notion of sense has at least three functions in Frege's semantics: (1) the sense of an expression is what speakers know when they understand it - this is an epistemic notion (sense as a cognitive value); (2)

the sense of an expression is the way of determining its reference - this is a semantic notion (sense as a method, route or criterion toward reference); and (3) the sense of an expression is its reference in oblique contexts (sense as the object of propositional attitudes). The problem that Frege's theory raises is, typically: How can the same thing play these different roles? In particular, how can the sense of an expression account for the individual speaker's competence and at the same time fix the reference of his words? Putnam's view is that there is no *single* notion that can honestly play roles (1) and (2) at the same time. In Putnam's theory role (1) is played by stereotypes, but stereotypes do not play role (2) as long as they underdetermine the meaning of natural-kind words. Now, Putnam's theory raises the opposite problem: How can we maintain a *unitary* concept of meaning if two different things account for the individual speaker's competence and for the determination of reference? If, as Putnam claims, the traditional concept of meaning "rests on false theory"[28] and "has *fallen to pieces*",[29] are we to take seriously Putnam's protests that he has come "to revise the notion of meaning, not to bury it".[30] Are there not reasons to believe that his semantic theory is another form of "Quinine scepticism in disguise"?

     If we take the second horn of the dilemma, there is a possible way cut that leaves assumption (II), that intension determines extension, unharmed. Suppose a supporter of Putnam's views were to say that the extension of 'water' is partially fixed by the following specification: *Water is whatever is identical in nature with THIS* (indexical identified) *stuff, whatever its nature may be.* There would be no objection to the idea that such a specification could partially fix the extension of the term 'water'. But, and this is all important, to adopt this idea would amount, in practice, to converting Putnam's semantic theory into a modified version of traditional Fregean semantics. For the notion of being identical in nature with this substance is nothing else but a part of the *concept* that fixes the reference of the word. That is, Putnam's theory then becomes a variety of traditional Fregean semantics which allows *indexical* conceptual contents. And as long as it is not proved that a Frege-like theory is radically unable to accommodate indexical conceptual contents, Putnam's claim that the traditional concept of meaning rests on a false theory is ungrounded.[31]

     Putnam counters this gambit in *Representation and Reality*. He claims, first of all, that indexical descriptions like 'stuff that behaves like and has the same composition as *this*' (used by someone who is "focusing" on a paradigm sample of the substance) cannot be synonymous with the terms whose reference they may help to fix. But we have seen that we can dispense with the requirement of synonymy. Putnam contends also that such an indexical description may be associated with qualitatively identical mental representations in the heads of an Earthian and an Twin-Earthian and still pick out different extensions, different substances - H_2O and XYZ:

> if the water I am focusing on looks and tastes just like the "water" that Twin Earth Hilary is focusing on, then my "mental representation" of my example may be "qualitatively" identical with Twin Earth Hilary's representation of his example. But the *stuff* is different, and so the *property* of being-pretty-much-like-*this* is a different property when I define it that way from the *property* of being-pretty-much-like-*this* when Twin Earth Hilary defines that way.[32]

Now this line of reasoning seems to suggest that indexical properties like being-pretty-much-like-*this* cannot be part of the conceptual content that fixes the reference. But this suggestion presupposes the identification of conceptual contents with mental representations such that two mental representations are qualitatively identical unless they are representations of superficially distinguishable items. Once again we meet a too narrow view of conceptual contents. If we adopt the more liberal view referred to above, the conceptual content associated with a term could include the stereotype and the indexical property. On the other hand, when Putnam asserts that the indexical property is different, since the stuff is different, he is taking again for granted that criteria of chemical composition predetermine the question of what counts as sameness and difference here.

VII. SOME PROBLEMS WITH THE NOTION OF ESSENCE

The specification *X is water only if X has the same nature as the paradigm instances, whatever that nature may be*, not only introduces an indexical content, it introduces also a *determinable* conceptual content. What counts as sameness of nature? Superficial features - Lockean nominal essences - do not determine identity of nature because a natural kind can have abnormal members. The presence of these features must be explained in terms of a

certain *essential nature*, Putnam contends. But then the question is posed
again with respect to the real essence: What counts as sameness of real
essence? Putnam is well aware that there is a problem here when he writes
that, in saying that we apply the word 'horse' to all things of the same kind,
"'of the same kind' makes no sense apart from a categorial system which
says what properties do and what properties do not count as similarities. In
some ways, after all, anything is 'of the same kind' as anything else".[33] In
practice, what Putnam does is to appeal to *structurally* important properties,
"the ones that specify what the liquid or solid, etc., is ultimately made out
(...) and how they are arranged or combined to produce the superficial
characteristics".[34] The "ultimate" atomic composition, the genetic code, the
DNA, etc., would be the properties which determine the essential nature of a
kind. Putnam's solution is to appeal to science. Things and kinds have
essences, but these essences are not determined - as Wittgenstein put it - by
grammar, but they are discovered by scientific investigation Since as a
matter of fact science has discovered that water is H_2O and since it was
within the speakers' *referential intentions*, even before this discovery, that a
liquid should *count* as water only if it had the same ultimate composition as
the paradigm samples of water, it follows that water is H_2O in every
possible world - or, what amounts to the same thing, that the statement
'Water is H_2O' is *metaphysically necessary.*

   This view does not avoid problems. To say that a statement is a
necessary truth is to say that it is true and cannot be false - that there are no
conceivable circumstances under which it is false. But then how can the
thesis that science discovers essences be conciliated with the fact that science
is *fallible*? If the statement 'Water is H_2O' gives us the essence of water,
then there is no possible world in which water is not H_2O. But why should
be inconceivable to count as water a substance in a possible world which
consisted of $H_{20}O_{10}$ molecules and which had macroscopic properties similar
to those of our water? Putnam deals with this difficulty in *Realism and
Reason*. His way out is to replace the notion of identity of essential
composition with that of *similarity to the paradigm*: "These examples
suggest that the 'essence' that physics discovers is better thought of as a sort
of *paradigm* that other applications of the concept (...) must *resemble* than as
a necessary and sufficient condition good in all possible worlds".[35] But this
way out has two consequences: (1) the notion of similarity to the paradigm
is vague and interest-relative, as it was pointed out above: and (2) this
solution damages the notion of metaphysical necessity: if what counts as

relevant similarity varies, the necessity of the statement 'Water is H_2O' vanishes.

Lastly, the idea that essence is also the product of the speakers' referential intentions poses certain difficulties too. The problem which Putnam tries to solve is the problem of the *projectibility* of meaning: How can we explain that the words 'horse' refers not only to the horses with which we have had some sort of casual contact, but also to all past, present and future horses? At first sight Putnam's answer seems clear: the natural kind itself plays a role in the determination of the reference of the word; the word 'horse' applies to all objects *of the same kind*. But here again we are caught in the same spiral as before. 'Of the same kind' means: sharing the same essential nature. But what counts as *sameness* of essence? It seems that there are only two possible answers to the question: either *we* are who determine - through intentions, conventions, rules, practices, forms of life, etc. - what counts as sameness of essences, or there are *objective essences* "out there", waiting to be discovered. And now Putnam's dilemma is: if he opts for the first solution, then I do not see how he can avoid one form or another of the Wittgensteinian view that essence is determined by grammar; but if he opts for the second way out, then he has no choice but to admit that there is after all a ready-made world, that there are Self-Identifying Objects. Putnam claims that there is a third way which allows us to assert both that there are objective essences and that these essences are not mind-independent. This third way would be internal realism. Whether it is a real solution or not is something that we shall examine another day.

NOTES

* A first version of this paper was presented in a colloquium on the philosophy of Hilary Putnam, held in Madrid in March, 1985. Successive drafts were read and criticized by my colleagues and nevertheless friends, Francisco Valle Arroyo, José Luis G. Escribano and Jorge Rodríguez Marqueze, but I alone am responsible for the possible remaining errors.

1 "Is Semantics Possible?", "Explanation and reference", and mainly "The meaning of 'Meaning'", all collected in Hilary Putnam, *Mind, Languages and Reality, Philosophical Papers*, vol. 2, Cambridge: Cambridge University Press, 1975 (quoted hereafter as MLR).

2 *MLR*, p. 222.

[3]   *MLR*, p. 219.

[4]   *MLR*, p, 199.

[5]   Cf. #3 of this paper.

[6]   Cf. note 5.

[7]   *MLR*, p. 274.

[8]   *MLR*, p. 229.

[9]   *MLR*, p. 245.

[10] *MLR*, p 234.

[11] *MLR*, p, 220.

[12] Ibid.

[13] *MLR*, p. 221 (italics and capital letters are mine).

[14] Cambridge, Mass.: The MIT Press, 1988.

[15] Op. cit., p. 29.

[16] And remember Putnam's liberal interpretation of 'psychological state' referred to in #1 (cf. notes 3 and 4).

[17] *Representation and Reality*, p. 29.

[18] *MLR*, p. 221.

[19] Tyler Burge, "Individualism and the Mental". in P.A. French, T.E. Uehling, and H.K. Wettstein, eds., *Midwest Studies in Philosophy*, vol. IV: *Studies in Metaphysics*, Minneapolis: University of Minnesota Press, 1979, pp. 73-121, and mainly "Other Bodies", in A. Woodfield, ed., *Thought and Object: Essays on Intentionality*, Oxford: Clarendon Press, 1982, pp. 97-120.

[20] *MLR*, p 221.

[21] *MLR*, p. 277.

[22] *MLR*, p. 239.

[23] Cambridge: Cambridge University Press, 1981. Ch.. 2: "A Problem about Reference".

[24] *Realism and Reason: Philosophical Papers*, vol. 3, Cambridge : University Press, 1983, Ch.. 3. "Possibility and Necessity".

[25] *MLR*, p. 151.

[26] *MLR*, p. 246.

[27] Cf. Michael Dummett, *Frege: Philosophy of Language*, London: Duckworth, 1973, pp. 91-95.

[28] *MLR*, p. 219.

[29] *Reason, Truth and History*, p. 29.

[30] *MLR*, p. 253. In fact, in *Representation and Reality*, Putnam avoids assumption (II) and replaces it with adequate versions of (C1) and (C2), taken now as two of the three assumptions on which a traditional semantic theory rests. (The remaining assumption has been stated in #3.)

[31] That the dependence of meaning (= intension) on causal relations does not pose greater problems than its dependence on *other* kinds of indexicality was pointed out by David Lewis in "Language and Languages" (in K. Guderson, ed., *Language, Mind and Knowledge*, Minneapolis: University of Minnesota Press, 1975), pp. 15-16.

[32] *Representation and Reality*, p. 33.

[33] *Reason, Truth and History*, p. 53.

[34] MLR, p. 239.
[35] *Realism and Reason*, p. 64.

Alfonso García Suárez
Universidad de Oviedo
Spain

JUAN J. ACERO

ATTITUDES, CONTENT AND IDENTITY:
A DYNAMIC VIEW

In this paper I shall be dealing with Frege's puzzle, which is a common topic nowadays within philosophy of language and mind. It amounts to the following question: how is it possible for anyone, e.g., Luke Skywalker, to understand and accept

(1)   Anakin Skywalker does not serve the emperor

without accepting - indeed, emphatically denying -

(2)   Darth Vader does not serve the emperor,

even though Anakin Skywalker is none other than Darth Vader? Also, how is it possible that whereas

(3)   Anakin Skywalker is Anakin Skywalker

is absolutely vacuous as its information value is nil,

(4)   Anakin Skywalker is Darth Vader

may contain the most fascinating discovery of a lifetime? A significant part of contemporary semantics and philosophy of language has centered on questions which differ from these only in the anecdotes that illustrate them. Though attempts to answer them have hardly ever taken into account that Frege's questions in fact pose a problem concerning the nature of our beliefs (and other mental states), it has usually been assumed that these and other mental states have a fixed and standing content. In this paper I shall suggest,

135

A. Clark et al. (eds.), Philosophy and Cognitive Science, 135–158.
© 1996 *Kluwer Academic Publishers.*

first, that a way out of this puzzle is to postulate that beliefs have non-stable, dynamic contents; and secondly, that such a solution as this is compatible both with what the New Theory of Reference holds within the philosophy of language and what Externalism maintains within the philosophy of mind.

I. STANDING CONTENTS

The puzzle posed by sentences (1) and (2) seems to be intimately linked to both the notion of belief and the idea of a belief having a content. If I believe that Anakin Skywalker does not serve the emperor, how can I possibly not believe that Darth Vader does not serve the emperor, provided that Anakin is Darth Vader? Well, here is one answer: I accept the former but not the second for I am far from sure that Anakin Skywalker is Darth Vader. (This possibility does not even cross my mind.) As soon as we decide to abide by the so-called New Theory of Reference (or New Theory of Mental Content),[1] we find that this answer is out of bounds. Those who conform to this theory consider that the content of the two beliefs is the same: if you believe that Anakin Skywalker does not serve the emperor, you believe that Darth Vader does not serve the emperor either; if you believe that Darth Vader serves the emperor, you also believe that Anakin does too. Quite simply, though, we are wrong in making out that we might believe one thing and not the other. What happens, we are told, is that "in certain cases one may be in no position to determine the consistency of one's statements and beliefs."[2]

Implicitly, this diagnosis goes hand in hand with the idea that the meaning of a sentence 'p', i.e. the content of the belief that p, is standing, something fixed and stable. Thus, from such a viewpoint sentence (1) expresses a *singular* proposition, that is, a structured entity to which individuals and properties can belong as single elements; not mental representations, but a complex made up of individuals and properties. To be exact, (1) expresses something like

(1a)   <SERVE, a, e>,

where **a** and **e** are respectively Anakin Skywalker in person, that is to say, Darth Vader, and the emperor in person. In addition to this,

(5)   Luke Skywalker believes that Anakin Skywalker
        does not serve the emperor

is understood as expressing a relationship between an agent, Luke Skywalker, and that very proposition or content:

(5a)   <BELIEVES, l, <SERVE, a, e>>.

For a New Theoretician of Reference, sentences (1) and (2) - as well as (3) and (4) - do not make any difference; and the same occurs with (5) and (6):

(6)   Luke Skywalker believes that Darth Vader does not
        serve the emperor

no matter what our linguistic and psychological intuitions might dictate.[3] Even though pairs of sentences like (1) and (2) or (5) and (6) are made out of different proper names, any difference in the propositions respectively expressed have to be accounted for in terms of differences in other constituents of the sentence. (It is assumed that any two proper names contribute in the same way to a proposition if they refer to the same individual.) Because of this, a New Theoretician uses a coarse net to filter out propositions or contents.

Opposition to this way of analysing (belief-)sentences is common in philosophical semantics. Critics seem to be fairly sure that accepting (1) cannot be exactly the same as accepting (2). When this line of argument is followed to its conclusion, it is pointed out that unless any two sentences differ in strictly synonymous constituents they are always bound to express different propositions. As supporters of the Language of Thought Hypothesis, for example, embrace this doctrine, they usually consider Frege's puzzle, far from being an obstacle, as a welcome confirmation of their own creed.[4] If propositional attitude sentences mean relationships between agents and representations and if these are symbolic forms of natural language or some other language of thought, beliefs expressed by (5) and (6) correspond to one and the same relationship C between agents and *different* sentences (or representations), no matter how we take it to be,:

(5b)   Luke Skywalker C 'Anakin-Skywalker-does-not-
            serve-the-emperor'
(6b)   Luke Skywalker C 'Darth-Vader-does-not-serve-
            the- emperor'.

As at the other end of the spectrum, subordinate clauses also have a fixed
meaning here:[5] each sentence has its own, right up to the very brink of
synonymy. Identification criteria turn out to be extremely fine. Since (1) and
(2) are different sentences and it is assumed that 'Anakin Skywalker' and
'Darth Vader' are not synonymous terms, (5) and (6) mean different beliefs,
i.e., beliefs with different contents.

Now then, do sentences (5) and (6) always describe the same beliefs
or not? No, they don't. If an agent A believes that a is P and at the same
time (actively) believes that $a =b$, then she will also believe that b is P. If A
believes that a is P, but does not believe that $a = b$, then she will not believe
that b is P. Luke does not believe that Darth Vader does not serve the
emperor, because he lacks the belief that Anakin Skywalker is Darth Vader,
though he accepts that Anakin Skywalker does not serve the emperor. The
Roman astronomer looking at Phosphorus in the morning does not believe
that he is looking at Hesperus because he does not believe that Hesperus is
Phosphorus. When new information arrives and new beliefs join older ones,
bringing about changes in the previous belief system, the same old sentence
can be used to ascribe a different content. Luke already knows who his father
is and believes that Darth Vader no longer serves the emperor. In such
circumstances, Luke believes that Anakin Skywalker does not serve the
emperor and that Anakin is Vader. Therefore, he also believes that Darth
Vader does not serve the emperor. Moreover, suppose that the Roman
astronomer has managed to find out that Hesperus is Phosphorus. When he
looks at Phosphorus in the morning and it occurs to him that Phosphorus is
hardly visible at that time of day, it will also occur to him that Hesperus is
hardly visible in the morning. The content of an agent's particular belief
heavily depends on further beliefs of hers, that is, on a substantial part of the
rest of her belief system. It is not true that (5) always expresses the same
belief as (6); nor is it true that (5) never expresses the same belief as (6).
What Luke Skywalker believes concerning who serves the emperor is a
function of further beliefs of his. A theory of meaning (or content) of
propositional attitude sentences must take into account not only Frege's data
but also this sort of context-dependency of subordinate clauses. Thus, the
problem for a theory of non-stable ascription is this: what form must it take?

II. FREGE'S PUZZLE: NOTIONAL WORLDS

I have coined my answer to the above question in the following terms. A belief sentence, i.e., a sentence of the form

(*) A believes that p

says that according to the way A sees the world, it is the case that p; in other words, if the world were as A thinks, then it would be the case that p. Let us call such a hypothetical world A's *notional world* and A's notional world is W. The question that really concerns me is this: if A's notional world is W, in what way does A's coming to believe that p potentially transforms W into a (possibly different) notional world W'? It is this transformation that fixes the content of the subordinate clause in (*) - which I shall call the B_A-*sentence* -, 'p', computing it by combining the linguistic meaning of 'p' and notional world W. This way of analysing how belief sentences is inspired by a *genetic* account of mental state contents. What one believes in a certain situation does not only depend on the situation one finds oneself in or the state of affairs one is linked or attuned to. It also depends on what else one actively believes at that moment.

　　　This genetic stance sheds some light, I feel, on what happens in the sort of situations which give rise to puzzles like Frege's. Luke believes that Anakin Skywalker does not serve the emperor without believing that Darth Vader does not serve him either, because in believing the former he has no reason to think that Anakin is Darth Vader. Similarly, Philip Marlowe believes that Terry Lennox died in a small Mexican village and that he is now talking to Cisco Maioranos, for he does not suspect that Lennox is Maioranos. Cases like these share the following traits. In the first place, an agent A believes that an individual a is P. Secondly, she also believes that a is not someone else, b. Thirdly, she is wrong about this last point: she ignores that a is b. Fourthly, the belief that a is not b arises out of the belief that the individual b has a property Q and the belief that there is nobody who is both P and Q (at the same time). The answer to the question whether A's belief that a is P is the same as her belief that b is P must be 'no'. A's notional world is such that there are at least two different individuals in it, one being P and the other not. And *this* notional world obviously differs from the one in which there is only one individual. One may accept sentence (1) without accepting (2), as (1) could be true in a notional world that would refute (2), and vice versa. Both sentences could partially describe the same notional world if Luke believed that Anakin Skywalker was Darth Vader. But this is not so. In a sense, Luke does not know who Darth Vader, (that is, his own father) is.[6]

A parallel comment applies to sentences (3) and (4). These are sentences which may be used to specify the mental content of an agent. For this purpose, the former sentence - not the latter - would be deadlock. Any notional world that can be described with the help of the proper name 'Anakin Skywalker' is one in which Anakin Skywalker is Anakin Skywalker, but not any world which can be described by means of 'Anakin Skywalker' and 'Darth Vader' is one in which both expressions are co-referential, or one in which two different communication chains end up in the same individual. To mark the difference between pairs of sentences such as (3) and (4), a famous term associated with the name of Frege - 'cognitive value' - is used. By means of this term it would be said, then, that these sentences differ in their cognitive value (or meaning). According to our way of tracing out such a difference, two expressions have different cognitive value if they do not potentially contribute the same thing to the description of the notional world of an agent.

Here 'potentially' is an adverb with an important role. Two expressions - i.e., two sentences - may have different cognitive values, but their contribution may be significantly similar in a certain case: when the agent believes that an individual a has a property P and also believes that $a = b$. According to this notional world she believes that b is P as well. So, although sentences 'a is P' and 'b is P' may each do their share on other occasions, in this case they make the same contribution. Or what amounts to the same thing, if agent A believes both that a is P and that $a = b$, then the sentence 'b is P' does not bring forward anything new. The meaning of a sentence may be standing, but when it comes to making a difference to an agent's notional world, it all depends on what her world is like.

In the semantic culture of the last two decades a distinction has been made between the *character* and the *content* of an expression and both have been recognized as aspects of its meaning. Character is a function from contexts of use to contents; contents are functions from possible worlds to semantic values, that is to say, extensions.[7] The character of the personal pronoun 'you', for instance, is a rule that assigns to every utterance of this pronoun its corresponding addressee(s). The content of any utterance of 'you' is the particular person or people referred to by it. In a similar vein, one would like to distinguish between the *cognitive meaning* of an expression and its *cognitive value*. The cognitive meaning is stable, a rule from notional worlds to notional worlds. The cognitive value is the specific modification which has been brought about: it is different things in different places. If you want to know the contribution to an agent's notional world M made by her acceptance of a sentence 'p', you must know beforehand what M is like and what 'p''s cognitive meaning is. A dynamic semantic theory of a B_A-*sentence* 'p' is one which determines the cognitive value of 'p' with regard to any notional world.

III. DYNAMIC SEMANTICS OF PROPER NAMES AND IDENTITY

My earlier remarks support the conception of a semantic theory of belief sentences as a kind of dynamic semantics of their B_A-sentences. In order to make the meaning of sentences of form (∗) explicit, we had better forget about 'A' and 'believes that' and restrict ourselves to saying what A's notional world would be like after incorporating 'p''s contribution into it. In speaking of a dynamic semantic theory of a class of sentences, I am following an idea which has been put forward on a number of occasions. In this approach, by the dynamic meaning of a sentence we should understand not its truth-conditions, but its *update*-conditions. If meaning is dynamic, then knowing the meaning of a sentence amounts to knowing how its acceptance by a speaker, or an agent, would potentially change her information state.[8] The reader can easily imagine what turn this approach is going to take in my hands: instead of characterizing an information state in the usual way - as a set of sentences, or sentences plus structural cues concerning sentences and relationships among them - I first identify an agent's information state with her notional world; and secondly I identify an agent's notional world with a (partial) model, in the common, set-theoretic sense of the term. Thus, I think it appropriate to envisage a notional world as a structure consisting of a non-empty set (i.e., the so-called domain or universe of discourse) and a map from primitive singular terms to individuals in the universe of discourse and from primitive general terms to sets of n-tuples of individuals belonging to that universe. Models built up in a certain way - the way this theory endeavours to capture - contain how the speaker (or the agent) A understands a text portraying the world she subscribes to. Seen in this light, a dynamic theory of B_A-sentences consists of a recursive definition of a transformation operation τ_A, a function mapping notional worlds to notional worlds. The argument of τ_A represents A's notional world as a context or parameter: that is, a belief system serving as a background against which a propositional attitude attribution may determine a content. The value of τ_A represents the notional world which emerges after incorporating the content of a new belief.

We can adopt this theoretical framework and use it to provide a systematic solution to Frege's puzzle. Such a solution requires τ_A definition clauses for both (i) proper names and (ii) the identity expression ('is' or 'is the same as'). The essentials of this solution are contained in the following two rules:

[PrNa]   If 'N' is a proper name, M is a notional world and 'N' is new with respect to M (i.e., the semantic value of 'N' is not defined in M), $\tau_A('N', M)$ is a new individual; otherwise, $\tau_A('N', M)$ is the individual previously assigned to 'N' in M.

[Id]   If 'N_1' and 'N_2' are singular terms and M is a notional world, $\tau_A \cdot (N_1 = N_2 \cdot)$ comprises (i) the addition of both $\tau_A('N_1', M)$ and $\tau_A('N_2', M)$; (ii) the shrinking of M's universe of discourse by getting rid of $\tau_A('N_1', M)$, and (iii) the systematic substitution in M of $\tau_A('N_1', M)$ by $\tau_A('N_2', M)$.

Rule [PrNa] puts forward two ideas: first, the notional contribution of a proper name 'N' is simply an individual: the individual called 'N'; and second, proper names that are new to an agent's notional world history bring forward new individuals. Thus, suppose that Philip Marlowe has never heard of anyone called 'Cisco Maioranos', as we are told in *The Long Goodbye*. On hearing an utterance of this name Marlowe reacts by adding a new individual to that part of his ontology made up of people, the new individual being a guy that Marlowe cannot locate among those previously known to him. On the other hand, when the proper name is not new to the agent, *its* conversational contribution will have already been made at a previous stage of his notional world history. To put it in terms of the mental archive, the same idea can be conveyed in the following way: a proper name is always linked to opening a new entry in a file which will contain every piece of information given by means of the proper name itself. As far as an old proper name is concerned, it works by ticking off information to be included in an existing file which is already open.

The second rule, [Id], tells us that the real contribution of an identity sentence may be the shrinking of the notional world acting as a context. The resulting model, M', will have the same universe as M minus one of its previous members. Since by accepting the identity $a = b$, we identify a as b, a simply becomes superfluous and we can get by quite well with b on its own. However, we still keep the name 'a' as a reminder that b is known by at least two different proper names. Thus, when the Roman astronomer comes to accept 'Hesperus is Phosphorus', he no longer distinguishes Hesperus from Phosphorus and his belief system reaches a new stage in which only one star has the many properties of both entities. (Some could be left out by modelling a new consistent system.) In this case, the notional contribution of an identity sentence with two proper names - which is the crux of Frege's puzzle - amounts to merging formerly unrelated files. (At least this is what characterizes the substantial, non-flimsy case.) A situation in which two files seem to contain information on internally different individuals changes into one in which several items of information about the same internal individual are contained in two files. Such cross-referring, commonly seen in encyclopaedias, means that more data on the same topic can be found elsewhere.[9]

Rules [PrNa] and [Id], as well the approach backing them up, smooth the way to showing how Frege's puzzle comes about and what can we do to solve it. The puzzle arises because on the one hand we know that Hesperus is Phosphorus (that $a = b$) and therefore that 'Phosphorus is visible in the morning' contributes to *our* unmentioned notional world in the same way as 'Hesperus is visible in the morning' does. However, we feel that there is something very specific to each sentence, since the Roman astronomer accepts the first and rejects the second. How can we resolve this conflict? The key to the conundrum lies in that different notional worlds - i.e., notional worlds with different universes - are involved in these two situations. While *ours* does not discriminate between a and b, the Roman astronomer's does.

Another point is worth discussing here. As is well known, Frege thought he had found in this conflict enough evidence to conclude that, apart from its referent, a singular term - even a proper name - has a sense apart from its reference. And this is a descriptive content enclosing a way of thinking about the term's referent. My diagnosis of Frege's puzzle allows us to avoid not only approving of the referent/sense distinction, but also negotiating the New Theory of Reference. According to [PrNa], a proper name contributes an individual to an agent's notional world, and that is all.

Exactly what individual it contributes depends on what the rest of this world is like. However, if you

(i)   restrict yourself to standing contents,
(ii)   respect the agent's psychology,

and

(iii)   choose not to opt for a descriptive theory on notional contribution of proper names,

then I am afraid that Frege's puzzle is an obstacle you cannot overcome. Frege thought that the way out of the puzzle was to give up (iii). The New Theoretician prefers to look for a solution by giving up (ii). He argues (in footnote 3) that speakers' intuitions can be interpreted as evidence of a kind of phenomenon that has hardly anything to do with semantic content. Frege's data urge us to accept (ii), I think, and there are a host of powerful arguments, essentially Kripke's, against embracing a descriptive theory of mental content. At the same time there are interesting proposals in support of a dynamic approach to semantic content. All considered, I would yield to the temptation of giving up (i). 'Difficult cases' are difficult because we usually forget about dynamic contents.

IV. KRIPKE'S PUZZLE

A puzzle relating to Frege's was proposed by Saul Kripke in the late seventies.[10] In launching it, Kripke attempted to block those criticisms aimed at the thesis that singular propositions may be contents of sentences like (1)-(4). In a nutshell, these critics pointed out that if (1) and (2) express the very same proposition, then we could infer (6) from (5) and (4): such an inference would simply be an instance of the substitutivity of identity principle. However, the criticism goes on, this principle fails in belief contexts. OK., replies Kripke, but this objection amounts to nothing as it is not obvious that a breakdown in a logical principle is involved at all in cases where someone believes that a is P, believes that b is not P and it turns out that $a = b$. We can arrive at what appears to be an instance of logical inconsistency of a belief system by taking a completely different route. It is unreasonable to blame unpalatable conclusions on substitutivity: "The reason does not lie in any specific fallacy in the argument [that exploits this principle] but rather in the nature of the realm being entered."[11] It is possible to obtain a very strong form of Frege's puzzle by turning to two apparently different principles entirely: the Disquotational Principle [= *DP*] and the Principle of Translation [= *PT*].

[*DP*]	If a normal English speaker, on reflection, sincerely assents to '*p*', then he believes that *p*.
[*PT*]	If a sentence in one language expresses a truth in that language, then any translation of it into any other language also expresses a truth.

Armed with these principles, Kripke confronts the following remarkable but perfectly plausible story. Pierre is a common French speaker, born and bred in France. Somebody has spoken to him in glowing terms about London, which he calls 'Londres'. Through this and other experiences, he comes sincerely and on reflection to say

(7)   Londres est jolie.

Later, Pierre is forced to go abroad and moves to a place called 'London' by its inhabitants, the very city he had heard about years before. Unfortunately for him, Pierre settles in an unattractive part of London, a suburb full of uneducated people. He does not like London at all. Like his neighbours, he never leaves this part of the city. One more thing to complete the story. Since his neighbours do not speak any French and he cannot afford to go to a language school, he has to learn English by the 'direct method', without the help of a phrase book or translation manual. He learns that he lives in a city whose name is 'London'. As soon as he manages to hold a conversation in English, he flatly declares

(8)   London is not pretty.

Let us now apply principles *DP* and *PT* to (7) and (8), respectively. From Pierre's utterance of (7), the circumstances around it and *DP*, we can infer (9):

(9)   Pierre croit que Londres est jolie.

And from his utterance of (8), circumstances around it and *DP* we are allowed
to infer (10):

(10)   Pierre believes that London is not pretty.

If we now apply *PT* to both (10) and (9) and translate (9) and (10) into
English, we finally obtain (10) and (11):

(11)   Pierre believes that London is pretty.

Facts in Pierre's story, that make up a consistent and almost truthful tale,
and principles *DP* and *PT* lead us to the conclusion that Pierre has
contradictory beliefs. However, this way of seeing things is difficult to
swallow. Since it seems that Pierre is a logically acute agent, how is it
possible for him to accept that London is pretty and deny that London is
pretty at the same time? This is *Kripke's puzzle*.

Is there any solution to it? Kripke thinks that Pierre's is a real
puzzle; that the details of Pierre's story, principles *DP* and *PT* and the final
consequences are all watertight; that once we enter this area, "our normal
practices of interpretation and attribution are subjected to the greatest
possible strain, perhaps to the point of breakdown."[12] This is shown by the
fact that, according to Kripke, there is no plausible answer to the question of
whether Pierre believes, or does not believe, that London is pretty. For my
part, I do think that a way out of Kripke's puzzle can be found: one of the
principles used to generate it is responsible for the conclusion. To be exact, I
suggest that the Principle of Translation is guilty; and that a dynamic
approach to Pierre's story reveals how remarkable *PT*'s effects are.

Let us consider how Pierre's notional world evolves in time. In his
vision of the world, Pierre comes to think that there are at least two different
cities, Londres and London (the city called 'Londres' and the city called
'London'). He does not know that the city called 'Londres' by many people
in the real world is precisely the city called 'London' by many people in the
real world. Pierre first believes that the city he calls 'Londres' is pretty; and
much later he learns that the city he calls 'London' is not pretty. He becomes

a 'London'-user after 'Londres' was in his repertoire and has no reason at all to suspect that both cities are one and the same. Having had no chance to think that London is the city he used to call 'Londres', he smoothly infers, or goes on thinking, that 'London' is the name of a city which he knew nothing about until then. Therefore, Pierre's relevant notional world contains two individuals, Londres (a) and London (b): a has the property P (the property of being pretty) whereas b has the property of being not-P. When we recount Pierre's notional world's *history* or describe its evolution, nothing out of turn occurs. A dynamic stance on sentences (9) and (10) confirms this point: Rule [PrNA] simply approves of this way of fitting a new piece of information into an old one. The key element in this description is the way several items of information (each belonging to a different historical stage) join together so as to make up the consistency of Pierre's belief system. However, this balance vanishes as soon we move from (9)-(10) to (10)-(11), i.e., when we make our mind up to abide by *PT*. Since there occur two proper names, 'Londres' and 'London', in (9) and (10), and only one, 'London', in the pair (10) and (11), we can expect this change to make a difference *vis-à-vis* PT. The reason is that one of the two occurrences of 'London' is of necessity not new. There is nothing we can do to avoid concluding that Pierre's notional world is a contradictory one. Since Pierre is at all events a logically sound chap, we must deduce that *PT* forces this unwanted conclusion upon us. Translation can turn the new into old hat.

It is important to recognize that this analysis of Kripke's puzzle, as well as that of Frege's, is not based on a descriptive theory of proper names. [PrNa] simply requires an individual to be a notional counterpart of a proper name. No descriptive condition is added to that counterpart. When you point to an individual a and stick to rule [PrNa], you do not require that a be included in such-and-such a set. However, what you *could* expect from a τ_A-clause for definite descriptions 'the x such that Q' is to have to supply a 'Q'-set, i.e., the semantic interpretation of predicate 'Q', with exactly one individual if there were no 'Q'-set or if it were empty. I infer from this difference that [PrNa] is compatible with the idea - upheld the New Theory of Reference - that proper names do not have any descriptive meaning. You may argue in favour of this thesis and at the same time maintain that sentences like (1) and (2), or (3) and (4), do not have the same cognitive meaning. What is more, [PrNa] is incompatible with the requirement of semantic innocence. Those who defend semantic innocence hold that any sentence expresses the same proposition both within an indirect context and without.[13] A dynamic approach to belief sentences sees clauses such as

'believes that', 'thinks that', 'expects that', and so forth as signals of a
(partial) notional world description. Apart from this, any B_A-sentence is like
any other. What matters is which text serves as a context, i.e., the traits of
the notional world acting as a parameter for that B_A-sentence contribution.
Because we simply cannot count on this parameter's weight and preserve
semantic innocence at the same time, we must opt for one or the other. If
there is anything to a dynamic view of language, then semantic innocence,
but not the New Theory of Reference, should be in trouble.

A nice extension of [PrNa] explanatory powers is provided by
identity sentences with two tokens of the same proper name. Suppose, for
instance, that you hear about a man called 'Paderewski' in two very different
contexts. One Paderewski is known to you for his great talent as a piano
player and there is also a Paderewski who has become famous as a Polish
political leader.[14] Suppose you have no reason to think that the artist and the
politician are one and the same person. Then the natural thing to infer is that
there are two different people involved, two men who happen to have the
same name, a not uncommon occurrence. If [PrNa] were applied not to name
types but to name tokens, this would account for notional worlds evolving
in this way.

V. THINKING THE SAME. TRIANGULATION FUNCTIONS

So far, I have attempted to uphold the idea that a dynamic approach to
propositional attitude sentences provides us with a natural solution to two
central problems they create. A different question is whether the
achievements of others approaches in this area may also be obtained by the
one I have been recommending. In particular, I will now vindicate the New
Theory of Reference when starting out from a dynamic point of view.

In a defence of singular propositions as plausible sentence meanings
and mental state contents, John Perry has written recently:

> "One reason we need singular propositions is to get at what
> we seek to preserve when we communicate with those who are
> in different contexts. Fregean thoughts won't do, and neither
> will meet truth-values. Another reason is to get at the
> structure of belief. Philosophers who are bothered by singular
> propositions often complain that individuals can't be 'inside
> the mind.' But of course the properties and relations that are
> constituents of 'general propositions' are no more in my

mind than individuals are. Minds evolve in a very
Strawsonian world, where the ability to reidentify
individuals, and to use information picked up in one
encounter to guide action in a later encounter is crucial. That
we can usefully describe minds by reference to individuals
they have acquired information about, and that our concepts
of belief and the other attitudes embody such a way of
describing minds, should not be specially perplexing. New
Theories are better suited for dealing with cognition than the
alternatives."[15]

Here, Perry puts forward two reasons in favour of the New Theory of
Reference. His first argument is that we need singular propositions to be able
to say, as a matter of fact or of interpretation, that two speakers, or two
agents, are saying or believing the same in a certain situation, even though
they might be using different linguistic means to express what they were
thinking. In such cases, contexts may differ while contents may not. Perry's
second argument is still more engaging. He thinks that our world is a
Strawsonian one, in which identification and reidentification are usual and
necessary practices. Information picked up in an encounter in our
environment is used later in another episode. Since it is individuals,
properties, states of affairs and events that one identifies (and reidentifies),
why not conclude that our mental state contents are made up of, or simply
are, pieces of this stuff? Why not classify mental states by using to what we
come across in these encounters? As I believe it is worthwhile commenting
on these two arguments of Perry's, I shall dwell on them here and in the next
and final section.

Let us begin with an example. María knows Octavio Paz well.
When Paz won the Nobel Prize for Literature, María thought that Paz's
literary merit had been justly valued; she now esteems him as writer of great
worth. Ana, though, knows nothing about Paz's writings or personality. She
catches a glimpse on TV of a man being awarded the Nobel Prize but she
does not even remember his name. She only supposes him to be a writer of
great worth. If we now assert both

(12)   María believes that Octavio Paz is a man of great
         worth

and

(13)   Ana believes that the winner of the Nobel Prize for
            Literature is a man of great worth,

in one sense we can rightly say that María and Ana think the same thing.
What is more, according to how we interpret their cases, we can identify
what they think with the very same singular proposition, <IS A MAN OF
GREAT WORTH, *o*>. It is even plausible to think that María's and Ana's
images of Paz have nothing in common. While María's is the result of
prolonged experience, Ana's is only a fleeting glimpse on a TV screen. If we
had to describe these two notional worlds, we would be hard pressed to find
anything they could share. Whether we like or not, however, there may be
situations in which we *could* put both beliefs into the same slot, and a
singular proposition is a useful way of doing it.

　　　　This decision is far from being arbitrary. It highlights the fact that
*there is no single rule we can abide by in describing an agent's notional
world.* When we turn to a belief sentence to say what an agent believes, we
can choose between two main strategies or cognitive styles. First, we can
stick to the agent's narrow psychology and try to be faithful to her desires
and beliefs, those mental states which allow us to make sense of her
behaviour. In this sense we say that Marlowe believes that Lennox died in a
small Mexican village, that Cisco Maioranos is a Mexican citizen who lives
in Mexico D.F., and that Lennox is not Maioranos. We also take this line
when we say that the Roman astronomer believes that Phosphorus, but not
Hesperus, is visible in the morning. This would be the strategy we would
reject, however, if we attributed to Anakin Skywalker the belief that Darth
Vader does not serve the emperor; and the same thing would happen if we
said that Ana believes that Octavio Paz is a man of great worth. Had we
opted for this strategy, we would not have been interested in the agent's
psychology but, to put it in David Lewis's terms, in "[her] dealings with the
things around [her], as happens if we are interested in [her] as a partner in
cooperative work and as a link in channels for information."[16] On these
lines, what matters is our agreement with the agent about what things are
like, by which we share the same individuals and properties. When we then
say that she believes or that he has found out that such-and-such, we feel
compelled to add some clause to explain how she believes or knows. We say,
for example, that she has discovered that such-and-such without knowing it,
and in this way we bring to light our own point of view *as attributers*.[17]

　　　　Each one of these general strategies gives rise to a form of
describing a notional world. In wanting to be faithful to a speaker's

individual psychology, we lay emphasis on those individuals and properties which meet her approval. She has the last word. If she utters 'the winner of the Nobel Prize for Literature is a man of great worth', and if she does so sincerely and on reflection, then we can make a guess at what her notional world is like in this regard: there must be exactly one individual who has (recently) won the Nobel Prize for Literature, and that individual has the property of being a man of great worth. What counts to get at her notional world is her behaviour, especially her linguistic behaviour - although this rule has to be qualified so as to resolve possible inconsistencies. In this respect, singular terms stand out as pointers to the speaker's ontology, that is, to what individuals she accepts; general terms she puts in her mouth announce what properties belong to individuals in her universe and what relationships exist among them; quantification phrases she uses indicate how her universe is classified into different subdomains, what these subdomains are like and what connections she traces out among them. On the other hand, we may want to keep our distance from the speaker's psychology and simply classify her beliefs and say what their contents are, by observing the trappings of that world. No psychology at all here, or the way she looks at the world, but what her links with it are. Here, our use of singular and general terms and quantification phrases is meant to help focus the objective world behind the screen of her mental states. It is important to stress that these two strategies - the *psychology* and the *agreement* strategies - do not necessarily exclude each other. Very often we *do* take advantage of both to get the best out of them: we describe what the world is like from the agent's vantage point and we do so by resorting to the world we share. Mental states are then classified by means of individuals, properties and relationships we know the agent has encountered in exploring the world around her. Since this is a very common case - the most common situation, I would say -, describing a notional world need not be systematically different from saying what situations and events the agent's mental state is causally linked to. It is the exceptions that call for analysis - and I am sure the New Theoretician would agree.

In an utterance of a belief sentence, then, two characters are involved, the utterer or speaker, S, and the agent, A, to whom a content is ascribed. S can either try to capture A's notional world, following the psychology strategy, or resort to the agreement strategy and link A to a situation whose ingredients are real-world individuals and properties. (Of course, he may combine the two.) In any case, two possibly different ontologies are involved, the agent's notional universe of discourse and real-world stock - represented by the speaker's unbiased voice. Taking them into consideration, it is up to us to admit the existence of what I shall call *triangulation functions*, which are (partial) functions correlating both S's and A's ontologies and sorting out what corresponds to what from each point of view. Thus, the *speaker's triangulation function* t_S tells us what real-world individuals correspond to the agent's notional individuals. Function t_S has notional individuals as arguments and, when defined, real-world individuals as values. Therefore, function t_S provides us with an answer to the following question: what real-world individuals are involved in A's beliefs? On the other hand, the *agent's triangulation function* t_A tells us what notional citizens correspond to what real-world individuals. t_A takes real-world individuals as arguments and notional individuals as values. Therefore, t_A specifies how A's experience of the world gives rise to the set of individuals that A will bet on.

        Triangulation functions explain a lot. When combined with a dynamic approach to belief sentences, they put within our reach what, according to Perry, singular propositions would win for us. Let us remember we need singular propositions to get at what is worth retaining when communicating with people who are in different contexts or when reporting mental states of subjects with different interests or perspectives. With speaker's triangulation functions we have the same content at hand for they fix that which is shared by two notional worlds from the speaker's point of view. They provide us with what we need when we say what two agents agree upon. Thus, in a way María and Ana agree about what Octavio Paz is like. When *we* say this, we are employing a triangulation function of ours which assigns the same values to possibly different individuals. In María's notional world, individual a (that is to say, Octavio Paz) is a man of great worth; in Ana's notional world, the winner of the Nobel Prize for Literature (that is to say, b) is a man of great worth. In saying that María and Ana believe the same, we describe their notional world with the help of a triangulation function such that $t_S(a) = t_A(b)$. Since, by hypothesis, $t_S(a) = t_A(b) = o$ (Octavio Paz), a singular proposition is determined in each case as the content of these beliefs. Thus, notional worlds plus speaker's triangulation functions give singular propositions.

        Another application of triangulation functions arises from Kripke's question as to whether Pierre believes or does not believe that London is pretty. As I have said, Kripke admits that there is no quick or direct answer to this question, and we can see why. In trying to say what Pierre believes, we must turn to his own notional world and, in order to get at it, we start out by checking what is the real counterpart of London. However, in this case t_A is not a function at all. Pierre's acquaintance with London does not univocally dictate what he is referring to. Since there are two Londons for Pierre, a and b, it so happens that both t_A(London) = a and that t_A(London) = b, but this is not enough to enable us to say with which notional city we have to identify the real London. What is more, Pierre's story is told in such a way that there is nothing to choose between a and b as the final value. Either option would be an arbitrary one. As a result, you cannot reasonably argue that Pierre's beliefs concern a and not b, or vice versa. There is no easy answer to Kripke's question. We must resort to exploiting metalinguistic resources, i.e., of saying that in Pierre's mind there is one city called 'Londres' and another called 'London'.[18]

VI. VERTICAL AND LATERAL RELATIONS: ROLES

Perry's second reason for embracing singular propositions as meanings of B_A-sentences is externalism: it happens to be adequate, given the nature of the mind, to describe its contents in terms of those individuals, properties and relations it comes across in reaching out to the world. This is a very powerful reason. By separating mind and world, the workings of the former would simply become a blur. Because of this, there seems to be no a priori reason to think that the project of describing the mind's contents in terms of individuals and another world's ingredients is doomed. An externalist conception of the mind cannot be wrong from beginning to end. I will devote the last section of this paper to showing how my dynamic approach to belief sentences and externalism might be reconciled.

Having accepted externalism, then, we are left with a doubt as to whether or not it contains the whole story about mental state content and above all about belief content. I would say not. I take beliefs to be cognitive particulars endowed with a content. This content is identified by the two kinds of relationships a mental state may enter into. On the one hand, there are causal relations between a mental state and those individuals, properties and situations present in its origin. Causal links, then, contribute much to build up the content of a mental state. Together with these *vertical* relationships, the sort of cases discussed in this paper persuades us to admit a second kind. Mental states are not only hooked on to the external world through causal relations; they are also connected to other mental states. As Donald Davidson has long insisted, beliefs do not exist in isolation but form blocks or subsystems making up richer mosaics, thanks to their links with perception, inference and action. Besides, according to Davidson, "the identity of a thought cannot be divorced from its place in the logical network of other thoughts, it cannot be relocated in the network without becoming a different thought."[19] Thus, for instance, since Philip Marlowe has Cisco Maioranos in front of him and since he believes that Lennox is dead, Marlowe addresses Maioranos the way he does - not the way one would talk to a dead man! I would not like to go as far as Davidson and claim that a belief content is determined exclusively by its relationships with other mental states, though I think this is often true. However, such a holistic or systemic touch is welcome in the analysis of puzzles like Frege's and Kripke's. What we have here are obvious instances of a mental state content being partially determined by other previous mental states. In order for a belief to become part of a system its content may be subjected to severe constraints and even forced to change somewhat. (If you do not believe that Terry Lennox is Cisco Maioranos and someone called 'Maioranos' is standing there and circumstances cooperate and you believe that Lennox is dead, you won't think that you are looking at Lennox.) This explains why the externalist factor is not enough. In believing that you are talking to Maioranos, you find yourself playing a game of "touch and go" in which vertical forces from the external world compete with lateral ones holding among mental states themselves. The real challenge is telling those cases in which vertical links prevail over those in which lateral connections win through.[20]

What happens in 'hard cases' is accounted for by carefully separating the externalist forces' contribution from the strain being put on the incoming information within the web of beliefs. An element of arbitrariness is inevitable in such an analysis, since the separation itself amounts to splitting data from theory - something, we are reminded, we have no objective basis to carry out. The alternative is giving up describing how the vectors involved determine the outcome. Externalist forces shed light on why the Roman astronomer believes that certain episodes are episodes of seeing one planet; or why they are one-planet-being-visible episodes. On their own, however, they do not suffice to settle identity questions, since they require a lot of theoretical background. Not having discovered that Hesperus is Phosphorus and having no reason to think that this is so, the Roman astronomer does not believe that Hesperus is visible in the morning; besides, he believes that Hesperus is not visible then. If we leave aside or do not know what his belief system is like, i.e., what lateral forces are being exerted inside it, we are on the brink of solving Frege's puzzle. These lateral forces fulfil a decisive role in fixing an agent's notional world; that is to say, in setting up her belief system content. In order to specify B_A-sentence content, we are bound to say which part of A's notional world this sentence describes. The point I am making here is that we cannot do so without taking into account the rest of A's belief system (or at least a substantial part of it).

Vertical relations, therefore, do not always have the last word. When it comes to an agent's narrow psychology, causal links sometimes may determine her narrow mental content. We have proof of the negative side in that, from a notional point of view, different causal encounters with the same individual - Anakin and Darth Vader, Hesperus and Phosphorus, Lennox and Maioranos, Londres and London - do not always count as meetings with the same individual. This is due to the fact that lateral forces determine what I shall call *roles*. Any notional individual embodies one role, its very own.[21] Each role is endowed with a number of properties. These properties give a profile, so to speak, to each role. I will say that an individual b called 'N' by A *occupies* the 'N'-role in A's notional world, if A thinks that b has each of those properties that she ascribes to a by using the name 'N'. 'Hard cases' are hard because the very same individual splits into two or more as different spatial, temporal or spatio-temporal stages of one real-world individual and these may end up occupying different roles. Thus, seen from the outside, different parts of London each play a specific role in Pierre's view of the world, i.e., the 'Londres' and the 'London'-roles, since each one is invested with its corresponding set of properties. Seen from the outside, two temporal cross-sections of Venus each play their own role in the Roman astronomer's notional world, i.e., the 'Hesperus'- and the 'Phosphorus'-roles. Seen from outside, two spatio-temporal cross-sections of Terry Lennox each play its own role in Marlowe's belief system, i.e., the 'Lennox'- and the 'Maioranos'-roles. However, when you see things from the agent's own position, there are no roles at all, only different individuals with their own personality.

In response to Perry's externalism, it is clear, then, that causal relationships help determine what real-world individuals occupy what roles. By this, I do not mean that causal relations do not contribute to fix content. Indeed, they very often do. Since in these cases the relation between causal origin and role is of a one-to-one kind, the individual causing a mental state with the content that a is P is the only occupant of the 'a'-role. Thus, from the existence of an 'N'-role we may easily and reasonably infer that the individual N exists. As a result, we think and talk about individuals who at the same time play certain roles and to whom we are related by virtue of

external links. This is not a hard and fast rule, though. You may have a role
with no occupant at all: in this case, you will think that there exists an
individual who is so-and-so and you will be wrong. A notional individual
does not necessarily have a real-world counterpart. When it has none,
notional and real worlds do not see eye to eye.[22]

NOTES

[1]   I am assuming that there is only one New Theory of Reference,
which is of course utterly false. I am claiming that this is so for the sake of
argument.

[2]   S. Soames, "Direct Reference, Propositional Attitudes, and
Semantic Content", in N. Salmon & S. Soames, eds.: *Propositions and Attitudes*
(Oxford University Press, 1988), p. 213.

[3]   This is not exact in every detail. A supporter of the New Theory of
Reference naturally expects to explain out Frege's data. However, she adds that
those data have nothing to do not with a theory of the *semantic content* of a
sentence but with a different one: a theory of the information conveyed by an
utterance of that sentence in a conversation (S. Soames, *loc. cit.*, #.5); a theory of
ways of conveying or interpreting such information (N. Salmon, *Frege's Puzzle*,
Cambridge, MA: The MIT. Press, 1986, cap. 8); or a theory of the information
created by an utterance of the sentence (J. Perry, "Cognitive Significance and New
Theories of Reference", *Noûs*, 22 (1988), 1 - 18, esp. p.8.)

[4]   The *locus classicus* is J. Fodor, "Propositional Attitudes", in *Re-
Presentations. Philosophical Essays on the Foundations of Cognitive Science*,
Brighton, Sussex: Harvester Press, 1981. Cfr.. also B. Loewer, "The Role of
'Conceptual Role Semantics'", *Notre Dame Journal of Formal Logic*, 23 (1982)
305 - 315, esp. p. 310f.

[5]   For a range of cases compatible with the form of
representationalism I have put forward, see W. Lycan, "Mental Content in
Linguistic Form", *Philosophical Studies*, 58 (1990) 147 - 154. My position in
the text corresponds approximately to the conjunction of theses (A), (B), (C), (D)
and/or (G) of Lycan's.

[6]   This should remind you of Jaakko Hintikka's way out of the wreck
of substitutivity of identity in indirect contexts in *Knowledge and Belief*. Cf. *op.
cit.* (Ithaca, NY: Cornell University Press, 1961), cap. 6. Cf. also "Quine on
Quantifying In: A Dialogue", *The Intentions of Intentionality and Other New
Models for Modalities*, Dordrecht: D. Reidel, 1975.

[7]   D. Kaplan, "Demonstratives, An Essay on the Semantics, Logic,
Metaphysics, and Epistemology of Demonstratives and Other Indexicals", in J.
Almog, J. Perry & H. K. Wettstein, eds.: *Themes from Kaplan*, Oxford University
Press, 1989; N. Salmon, *Frege's Puzzle, loc. cit.*, chap. 2 and Appendix C.

[8] The first suggestion in this sense was made by H. Kamp and incorporated in his Discourse Representation Theory [= DRT]. For a paper on the intuitions behind this theory, see "Content, Thought and Communication", *Proceedings of the Aristotelian Society*, vol. LXXXV (1984/1985) 238 - 261. One important idea is this: "Not only do the utterances we produce, orally or in writing, often depend on context for their interpretation; they also do much to determine what the context *is*" (p. 240). My proposal of analysing cognitive meanings as functions from notional worlds to notional worlds benefits from Kamp's idea. I do prefer to handle notional structures directly, since DRT depends heavily on a very strong form of representationalism, which explicitly welcomes a language of thought hypothesis. Because of this, several philosophers and semanticians are exploring a different form of dynamic semantics, one which does not commit itself to representationalism. See, for example, J. Groenendijk & M. Stokhof, "Dynamic Predicate Logic", forthcoming *Journal of Philosophical Logic* vol. 14 (1991) 39 - 100; "Dynamic Montague Grammar", *ITLI Prepublication Series*, University of Amsterdam, 1990; F. Veltman, "Semántica de actualización", *Revista española de filosofía*, no. 6 (1992); "Defaults in Update Semantics", *ITLI Prepublication Series*, University of Amsterdam, 1991.

[9] The distinction between *internal* and *external* reference to an individual by a (mental or extramental) file is put forward in J. Perry's "Cognitive Significance and New Theories of Reference", *loc. cit.*, p. 13.

[10] A Puzzle About Belief", A. Margalit, ed.: *Meaning and Use*, Dordrecht: D. Reidel, 1979. Also in N. Salmon & S. Soames, eds.: *Propositions and Attitudes, loc. cit.*

[11] "A Puzzle About Belief", *loc. cit.*, p. 268.

[12] Op. cit., p. 269.

[13] J. Barwise & J. Perry, *Propositions and Attitudes* (Cambridge, MA: The MIT. Press, 1982), chap. 8.

[14] I am alluding to one sort of case brought up by Kripke. See *Op. cit., #. III.*

[15] "Cognitive Significance and New Theories of Reference", *loc. cit.*, p. 4.

[16] D. Lewis, *On the Plurality of Worlds* (Oxford: Basil Blackwell, 1986), p. 59.

[17] See Gerald Weissmann's comments on Edmund Stone's findings in "Aspirine", *Scientific American*, March, 1991.

[18] This one of the reasons why metalinguistic statements of the form '*N* is called "*N*"' are so useful. See M. J. Wreen, "Socrates is called 'Socrates'", *Linguistics and Philosophy*, vol. 12 (1989), 359 - 371.

[19] "Rational Animals", *Dialectica*, vol. 36 (1982) 318 - 327. See also "Thought and Talk", *Essays on Truth and Interpretation* (Oxford University Press, 1984), p. 157. This theoretical position is known as Meaning Holism. For a critical analysis, see J. Fodor, *Psychosemantics* (Cambridge, MA: The MIT. Press, 1987), chapter 3.

[20] Adjectives such as 'vertical' and 'lateral' have been introduced in this connection by Brian Loar, to illustrate the point I make in the main text. See

B. Loar, *Mind and Meaning* (Cambridge University Press, 1981), chap. 4 and p.
121. Others oppose 'internal' structural properties to external relations. Cf. M.
Crimmins & J. Perry, "The Prince and the Phone Booth: Reporting Puzzling
Beliefs", *Journal of Philosophy*, vol. LXXXVI (1989), p. 687. As can be seen, I
opt for a two-factor theory of meaning and mental content. For recent defences of
such a position, see N. Block, "Advertisement for a Semantics for Psychology",
in P. A. French, Th. E. Uehling & H. K. Wettstein, eds.: *Midwest Studies in
Philosophy 10*, Minneapolis: University of Minnesota Press, 1986; J. Fodor,
Psychosemantics loc. cit., chaps. 3 & 4; F. Dretske, *Explaining Behaviour*
(Cambridge, MA: The MIT. Press, 1988), #. 6.4; C. McGinn, *Mental Content*
(Oxford: Basil Blackwell, 1989), chapters 2 (section entitled "Dual
Componency") and 3 (section "Holism"); P. Jacob, "Thoughts and Belief
Ascriptions", *Mind and Language*, vol. 2 (1987) #. V; "Externalism Revisited: Is
There Such a Thing as Narrow Content?", *Philosophical Studies*, **60** (1990) 143 -
176.

[21]   This idea is inspired by a comment made by R. Stalnaker in
"Indexical Belief", *Synthese*, **49** (1981) 129 - 151, esp. p. 139.

[22]   I am very grateful to Jean Stephenson who has steered me away
from many errors in my English draft.

Universidad de Granada
Spain

PETER GÄRDENFORS

CONCEPTUAL SPACES AS A BASIS FOR COGNITIVE SEMANTICS

I. PROGRAM

As an introduction, I want to contrast two general traditions in semantics, one *realistic* and one *cognitive*. According to the realistic approach to semantics the meaning of an expression is something out there *in the world*. In technical terms, a *semantics* for a language is defined as a mapping from the syntactic structures to things in the world (or in several possible worlds). Often meanings are defined in terms of *truth conditions*. A consequence of this approach is that the meaning of an expression is independent of how individual users understand it.

The second paradigm of semantics is conceptualistic or cognitivistic. The central tenet of this approach is that meanings of expressions are *mental entities*. A semantics is seen as a mapping from the linguistic constituents to *cognitive structures*. The external world enters the scene only when the relation between it and the cognitive structure is considered. According to this kind of semantic theory the relation between meanings and the external world is secondary, and only determined after the cognitive structures have been settled. As a consequence, meaning becomes independent of truth.

In this paper, I first give a sketch of a realist semantics in the form of standard intensional semantics, which is formulated in terms of possible worlds. I shall also mention some of the philosophical problems this kind of semantics leads to. Then I present some of the main tenets of what has become known as cognitive semantics. As an ontological framework for a cognitive semantics, I introduce the notion of a *conceptual space* and show how such spaces can be used as a basis for a cognitive semantics. In the final sections, it will be argued that this kind of semantics is useful for understanding *metaphors* and *prototype effects* of concepts.[1]

II. INTENSIONAL SEMANTICS AND ITS PROBLEMS

A typical example of a realistic semantic theory is the so called *intensional semantics*. As an analysis of natural language it reaches its peak with

159

A. Clark et al. (eds.), *Philosophy and Cognitive Science*, 159–180.
© 1996 *Kluwer Academic Publishers*.

Montague (1974). Here a language is mapped onto a *set of possible worlds*.
Apart from truth values, possible worlds and their associated sets of
individuals are the only primitive semantical elements of the model theory.
Other semantical notions are defined as *functions* on individuals and possible
worlds. For example, a *proposition* is defined as a function from possible
worlds to truth values. Such a function thus determines the *set of worlds*
where the proposition is true. According to traditional intensional semantics,
this is all there is to say about the meaning of a proposition.

As a typical example of the analysis within this tradition let us look
at the notion of a *property*. In intensional semantics, a property is something
that relates individuals to possible worlds. In general terms, a property can be
seen as a many-many relation P between individuals and possible worlds
such that iPw holds just when individual i has the property in world w. Such
a relation is illustrated in Figure 1.

In intensional semantics, functions are preferred to many-many
relations. There are two ways of turning the relation P into a function:
Firstly, it may be described as a *propositional function*, i.e. *a function from
individuals to propositions*. Since a proposition is identified with a set of
possible worlds, this means that a property is a rule which for each
individual determines a corresponding set of possible worlds. But we can also
turn the table around to get an equivalent function out of P: for each possible
world w, a property will determine a set of individuals which has w as an
element of the sets of possible worlds the individuals are assigned (cf. Figure
1). This means that an equivalent definition of a property is that it is *a
function from possible worlds to sets of individuals*.

In Gärdenfors (1991b), it is argued that the standard definition of a
property within intensional semantics leads to a number of serious problems.
First of all, the definition is highly counterintuitive since properties become
very abstract things. The definition is certainly not helpful for cognitive
psychologists who try to explain what happens when a person *perceives* that
two objects have the same property in common or, for example, why certain
colors look *similar*.

A related, but more serious, problem for the traditional definition of
a property is that it can hardly account for *inductive reasoning*. An inductive
inference generally consists in *connecting* two properties to each other. This
connection is obtained from a number of instances of individuals exhibiting
the relevant properties. If a property is defined as a function from possible
worlds to set of individuals, then in order to determine which properties are
instantiated by a particular individual (or a set of individuals), one has to

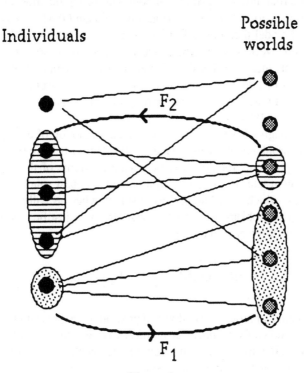

Figure 1.
A property as a many-many relation between
individuals and possible worlds.

determine *which functions* have the individual (or the set of individuals) as value in the actual world. Apart from problems concerning how we determine which is the actual world, this recipe will in general give us *too many* properties. For example, if we are examining a particular emerald it will instantiate a large class of Goodman-type properties like 'grue' apart from standard properties like 'green'. If the only thing we know about properties is that they are some kind of abstract functions, then we have no way of distinguishing natural and inductively projectible properties like 'green' from inductively useless properties like 'grue'. What is needed is a criterion for separating the projectible sheep from the non-projectible goats. However, classical intensional semantics does not provide us with such a criterion.[2]

The final problem that I shall point out for the functional definition of properties is perhaps the most serious one. Putnam (1981) has shown that the standard model-theoretic definition of 'property' which has been given here does not work as a theory of the *meaning* of properties. Putnam concludes that "there are always inifinitely many different interpretations of the predicates of a language which assign the 'correct' truth-values to the sentences in all possible worlds, *no matter how these 'correct' truth-values are singled out*" (1981, p. 35).

Although I have here only mentioned the notion of a property and some of the problems the realist semantics leads to, I believe that there are many other problems for this type of semantics. Some of these problems will be apparent later in this paper. But perhaps the best source is Lakoff's book (1987), which is a lengthy criticism of what he calls "objectivist semantics". The problems for the traditional semantics justifies a search for a fundamentally different kind of semantics.

III. COGNITIVE SEMANTICS

As an alternative approach, I shall give a programmatic presentation of what has become known as *cognitive semantics*. My presentation will be in the form of six slogans, with some comments, where the approach of a cognitively oriented semantics will be contrasted with the more traditional view. Prime examples of works in this tradition are Lakoff's (1987) and Langacker's (1986). Related versions of cognitive semantics can be found in the writings of Jackendoff (1983, 1990), Johnson-Laird (1983), Fauconnier (1985), Talmy (1988), Sweetser (1990) and many others. There is also a French semiotic tradition, exemplified by Desclés (1985) and Petitot-Cocorda (1985), which shares many features with the American (mainly Californian) group.

I.   Meaning is *conceptualization* in a cognitive
           model (not truth conditions in possible worlds).

The prime slogan for cognitive semantics is: *Meanings are in the head*. More precisely, a semantics for a language is seen as a mapping from the expressions of the language to some cognitive or mental entities. A consequence of the cognitivist position that puts it in conflict with many other semantic theories is that no form of truth conditions of an expression is necessary to determine its meaning. The truth of expressions is considered to be secondary since truth concerns the relation between a cognitive structure and the world. To put it tersely: *Meaning comes before truth.*

Cognitive semantics should be separated from Fodor's (1981) "Language of Thought" hypothesis. There are similarities, though: Fodor also uses mental entities to represent linguistic information. This is his 'language of thought' which is sometimes also called 'Mentalese'. According to Fodor, this is what speakers use when they compute inferences (according to some internal set of rules) and when they formulate linguistic responses (translated back from Mentalese to some appropriate natural language). However, the mental entities constituting Mentalese form a *language* with syntactic structures goverened by some recursive set of rules. And when it comes to the *semantics* of Mentalese, Fodor still is a realist and relies on references in the external world as well as truth conditions.

II.   Cognitive models are mainly *perceptually* determined (meaning is not independent of perception).

Since the cognitive structures in our heads are connected to our perceptual mechanisms, directly or indirectly, it follows that *meanings are*, at least partly, *perceptually grounded*. This, again, is in contrast to traditional realist versions of semantics which claim that since meaning is a mapping between the language and the external world (or several worlds), meaning has nothing to do with perception.

We can talk about what we see and hear. Conversely, we can create pictures, mental or real, of what we read or listen to. This means that we can translate between the visual form of representation and the linguistic code.[3] A central hypothesis of cognitive semantics is that the way we store perceptions in our memories has the *same form* as the meanings of words. Another consequence of the coupling of perceptual representation and meaning is that meaning has *ecological validity*.

III.   Semantic elements are based on *spatial* or *topological* objects (not symbols that can be concatenated according to some system of rules).

In contrast to the Mentalese of Fodor and others, the mental structures applied in cognitive semantics *are* the meanings of the linguistic idioms; there is no further step of translating conceptual structure to something outside the mind. Furthermore, instead of being a symbolic system having syntactic structure like Mentalese, the conceptual schemes that are used to represent meanings are often based on *geometric* or *spatial* constructions.

The most important semantic structure in cognitive semantics is that of an *image schema*. Image schemas have an inherent spatial structure. Lakoff (1987) and Johnson (1987) argue that schemas such as 'container', 'source-path-goal' and 'link' are among the most fundamental carriers of meaning. They also claim that most image schemas are closely connected to *kinesthetic* experiences.

> IV.   Cognitive models are primarily *image-schematic*
>   (not propositional). Image-schemas are trans-
>   formed by *metaphoric* and *metonymic* operations
>   (which are treated as exceptional features on the
>   traditional view).

Metaphors and metonymies have been notoriously difficult to handle within realist semantic theories. In these theories these linguistic figures have been treated as a deviant phenomenon that has been ignored or incorporated via special stylistic rules. In contrast, they are given key positions within cognitive semantics. Not only poetic metaphors but also everyday 'dead' metaphors are seen as central semantic features and are given systematic analyses. One of the first works in this area was Lakoff and Johnson (1980).

Metaphors and metonymies are primarily seen as *cognitive* operations, and their linguistic expression is only a secondary phenomenon. They are analysed as *transformations* of image schemas. As such they are connected to spatial codings of information. In particular, Lakoff (1987, p. 283) puts forward what he calls the '*spatialization of form hypothesis*' which says that conceptual forms are understood in terms of spatial image schemas plus a metaphorical mapping. For example, many uses of prepositions are seen as metaphorical (see e.g. Brugman (1981) and Herskovits (1986))

> V.   *Semantics* is primary to syntax and partly
>   determines it (syntax cannot be described
>   independently of semantics).

This thesis is anathema to the Chomskyan tradition within linguistics. Within Chomsky's school, grammar is a *formal calculus*, which can be described via a system of rules, where the rules are formulated independently of the meaning of the linguistic expressions. Semantics is something that is added, as a secondary independent feature, to the grammatical rule system. Similar claims are made for pragmatic aspects of language.

Within cognitive linguistics, semantics is the primary component (which, in the form of perceptual representations, existed before language was fully developed). The structure of the semantic schemas put constraints on the possible grammars that can be used to represent those schemas. To give a trivial example of how semantics determines syntax, consider the role of *tenses*. In a Western culture where time is conceived of as a line, it is meaningful to talk about three basic kinds of time: past, present and future. This is reflected in the grammar of tenses in most languages. However, in cultures where time has a circular structure, or where time cannot be given any spatial structure at all, it is not *possible* to make a distinction between, say, past and future. And there are languages which have radically different tense structures, which reflect a different underlying conceptual structuring of time.

VI.   Concepts show *prototype* effects (instead of following the Aristotelian paradigm based on necessary and sufficient conditions).

The classical account of concepts within philosophy is Aristotle's theory of *necessary and sufficient conditions*.[4] His view on how concepts are determined has had an enormous influence throughout the history of philosophy. During this century the Aristotelian notions became part of the program of the logical positivists who demanded that all scientific concepts should ideally be *defined* in terms of a limited number of observational terms. If a concept can't be defined by necessary and sufficient conditions, it is not a proper scientific concept, at least according to the early positivist program.

However, one very often encounters problems when trying to apply the Aristotelian theory. As a result of a growing dissatisfaction with the classical theory of concept theory, an alternative theory was developed within cognitive psychology. This is the called *prototype theory* where Eleanor Rosch is one of the main proponents.[5] The main idea of prototype theory is that within a category of objects, like those instantiating a property, certain members are judged to be more representative of the category than others. For example, robins are judged to be more representative of the category 'bird' than are ravens, penguins and emus; and desk chairs are more typical instances of the category 'chair' than rocking chairs, deck-chairs, and beanbag chairs. The most representative members of a category are called *prototypical* members.

Another thesis of prototype theory is that categories are not organized just in terms of simple taxonomic hierarchies. Instead, a 'middle' kind of concepts can be distinguished, which is called the *basic level* of the categorization. Higher levels are called *superordinate* and lower *subordinate*. For example, 'chair' and 'dog' are basic level concepts, while 'furniture' and 'mammal' are superordinate concepts and 'armchair' and 'dachshund' are subordinate. The basic level is characterized by a number of features: (1) It is the highest level at which category member have similarly preceived overall *shapes*, (2) it is the highest level at which a person uses similar actions for handling category members, (3) it is the level at which subjects are fastest at identifying category members, and (4) it is the first level named and understood by children.

Within cognitive semantics, one attempts to account for prototype effects of concepts. A concept is often represented in the form of an image schema and such schemas can show variations just like birds and chairs. This kind of phenomenon is extremely difficult to model using traditional symbolic structures.

IV. CONCEPTUAL SPACES AS A FRAMEWORK FOR A COGNITIVE SEMANTICS

After this presentation of some of the central tenets of cognitive semantics, I now want to turn to the *ontology* of such a semantics. As a framework for a cognitive structure used in describing a semantics I want to put forward the notion of a *conceptual space*. A conceptual space consists of a number of *quality dimensions*. As examples of quality dimensions let me mention color, pitch, temperature, weight, and the three ordinary spatial dimensions. The dimensions are taken to be cognitive and infra-linguistic in the sense that we can represent the qualities of objects without presuming an internal language in which these qualities are expressed. Some of the dimensions are closely related to what is produced by our sensory receptors, but there are also quality dimensions that are of an abstract non-sensory character.

The notion of a *dimension* should be understood literally. It is assumed that each of the quality dimensions is endowed with certain topological or metric structures. For example, 'time' is a one-dimensional structure which we conceive of as being isomorphic to the line of real numbers. Similarly, 'weight' is one-dimensional with a zero point, isomorphic to the half-line of non-negative numbers. Some quality dimensions have a *discrete* structure, i.e., they merely divide objects into classes, e.g., the sex of an individual.

A psychologically interesting example of a quality dimension concerns *color perception*. In brief, our cognitive representation of colors can be described by three dimensions (see Figure 2). The first dimension is *hue*, which is represented by the familiar *color circle*. The topological structure of this dimension is thus different from the quality dimensions representing time or weight which are isomorphic to the real line. The second psychological dimensions of color is *saturation*, which ranges from gray to increasingly greater intensities. This dimension is isomorphic to an interval of the real line. The third dimension is *brightness* which varies from white to black and is thus a linear dimension with end points. Together these three dimensions, one with circular structure and two with linear, make up the color space which is a subspace of our perceptual conceptual space.

I cannot provide a complete list of the quality dimensions involved in our conceptual spaces. Some of the dimensions seem to be *innate* and to some extent hardwired in our nervous system, as for example color, pitch, and probably also ordinary space. Other dimensions are presumably *learned*. Learning new concepts often involves expanding one's conceptual space with new quality dimensions. *Functional* properties used for describing artifacts may be an example here. Still other dimensions may be *culturally* dependent. 'Time' is a good example – in contrast to our linear conception of time, some cultures conceive of time as circular so that the world keeps returning to the same point in time, and in other cultures it is hardly meaningful at all to speak of time as a dimension. Finally, some quality dimensions are introduced by *science*.

This concludes my general presentation of conceptual spaces.[6] There is a strong similarity between the notion of a conceptual space and the *domains* as used in Langacker's (1986) semantic theory. The following quotation from Langacker (1986, p. 5) concerning his notion of 'domains' strongly supports this thesis:

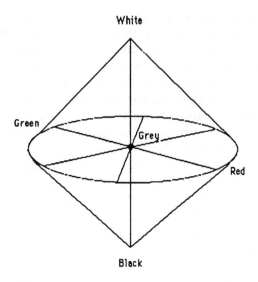

Figure 2
The full color space

"What occupies the lowest level in conceptual hierarchies? I
am neutral in regard to the possible existence of conceptual
primitives. It is however necessary to posit a number of 'basic
domains,' that is, cognitively irreducible representational
spaces or fields of conceptual potential. Among these basic
domains are the experience of time and our capacity for dealing
with two- and three-dimensional spatial configurations. There
are basic domains associated with various senses: color space
(an array of possible color sensations), coordinated with the
extension of the visual field; the pitch scale; a range of
possible temperature sensations (coordinated with positions
on the body); and so on. Emotive domains must also be
assumed. It is possible that certain linguistic predications are
characterized solely in relation to one or more basic domains,
for example time for (BEFORE), color space for (RED), or time
and the pitch scale for (BEEP). However, most expressions
pertain to higher levels of conceptual organization and
presuppose nonbasic domains for their semantic
characterization."

The theory of conceptual spaces is a *theory for representing information*, not
an empirical psychological or neurological theory, which I believe can be

applied to a number of philosophical problems in epistemology and semantics. Here, my primary aim is to show its viability as a foundation for cognitive semantics.

V. COGNITIVE SEMANTICS BASED ON CONCEPTUAL SPACES

I can only outline the first steps in developing a cognitive semantics based on conceptual spaces. According to the cognitive view, semantics is a relation between langauge and a cognitive structure. I submit that the appropriate framework for the cognitive structure is a conceptual space. This means that formulating a semantics for a specific language is to specify the mapping between the lexicon of the language and a conceptual space and to describe the operations on the image schemas defined on the conceptual corresponding to syntactic formation rules.

Slightly more technically, we can define an *interpretation* for a language L as a mapping of the components of L onto a conceptual space. As a first element of such a mapping, *individual names* are assigned vectors (i.e., points in the conceptual space) or partial vectors (i.e., points with some arguments undetermined). In this way, each name (referring to an individual) is allocated a specific color, spatial position, weight, temperature, etc.[7] If a name is assigned a partial vector, this means that not all the properties are known or have been determined. Following Stalnaker (1981, p. 347), a function which maps the individuals into a conceptual space will be called a *location function*.

As a second element of the interpretation mapping, the *predicates* of the language that denote primary properties are assigned regions in the conceptual space. (In Gärdenfors 1990, 1991 it is argued that the regions correponding to natural predicates are *convex*.) Such a predicate is *satisfied* by an individual just in case the location function locates the individual at one of the points included in the region assigned to a predicate. Some of the so called intensional predicates, like 'tall', 'former' or 'alleged', do not denote primary properties in the sense that their regions can be described independently of other properties. Such secondary predicates, which are 'parasitical' on other properties, can be described in terms of the regions assigned to the primary properties. *Relations* (primary and secondary) can be treated in a similar way.

If we assume that an individual is completely determined by its set of properties, then all points in the conceptual space can be taken to represent *possible individuals.* On this account, a possible individual is a *cognitive* notion that need not have any form of reference in the external world. This construction will avoid many of the problems that have plagued other philosophical accounts of possible individuals. A point in a conceptual space will always have an internally consistent set of properties – since, for example, 'blue' and 'yellow' are disjoint properties in the color space, it is not possible that any individual will be both blue and yellow (all over). There is *no need for meaning postulates* or their ilk in order to exclude such contradictory properties.

One important contrast to the traditional intensional semantics is that the one outlined here does not presume the concept of a *possible world*. However, different location functions describe alternative ways that individuals may be located in a conceptual space. Thus, these location functions have the same role as possible worlds in the traditional semantics. This means that we can *define* the notion of a possible world as a possible location function, and this can be done without introducing any new semantical primitives to the theory.

If we assume that the meanings of the predicates, among other things in a language L, are determined by a mapping into a conceptual space S, it follows from the topological structure of different quality dimensions that certain statements will become *analytically* true (in the sense that they are independent of empirical considerations). For example the fact that comparative relations like 'earlier than' are *transitive* follows from the linear structure of the time dimension and is thus an analytic feature of this relation (analytic-in-S, that is). Similarly, it is analytic that everything that is green is colored (since 'green' refers to a region of the color space) and that nothing is both red and green. Analytic-in-S is thus defined on the basis of the topological and metric structure of the conceptual space S. A consequence of this definition is that an analytic statement will be satisfied for all location functions. However, different conceptual spaces will yield different notions of analyticity.

With the aid of the notion of a conceptual space, I have tried to show how the slogans of cognitive semantics presented in Section 3 can be given some substantial content. That meaning is conceptualization (slogan I) is pretty obvious, given that the framework of a semantics is a conceptual space. Since many of the dimension of a conceptual space are directly connected to preceptual mechanisms, this also shows that the relevant cognitive models are perceptually determined (slogan II). And conceptual spaces are, by definition spatial, and not symbolic (slogan III). Of course, developing the mappings from an actual natural language to a cognitively realistic conceptual space is a Herculean task. Some first steps towards the completion of this task has been taken by the linguists in this tradition, as e.g. in the works by Langacker (1986), Lakoff (1987), Jackendoff (1990), and Talmy (1988).

VI. METAPHORS

As a way of filling out slogan IV within a cognitive semantics based on conceptual spaces, let us look at the way metaphors work. This is a problem which has been notoriously difficult to handle within realist semantic theories. In these theories, metaphors have been treated as a deviant phenomenon that should be ignored or incorporated via special rules. The view within cognitive semantics is that metaphors should be treated on par with all other semantic processes, or perhaps even as one of the central semantic features of language. Here I will present a summary of the theory of metaphors outlined in Gärdenfors (1992).

The core hypothesis is that *a metaphor expresses a similarity in topological or metrical structure between different quality dimensions.* A word that represents a particular structure in one quality dimension can be used as a metaphor to express a similar structure about another dimension. In this way one can account for how a metaphor can *transfer knowledge* about one conceptual dimension to another.

As a simple example, let us consider words that refer to the length dimension, like 'longer', 'distant', 'in front of', and 'forward'. This dimension refers to the most salient direction of the two-dimensional surface we are normally moving on. Unless altered by the communicative context, the default direction of this dimension is determined by the speaker's front and back. The spatial length dimension is represented by a topological structure that is isomorphic to the real line, where we, for the present purposes, can take the zero point to represent 'here'.

In our conceptual space (modern Western), the time dimension has the same structure as the real line. According to the hypothesis about how metaphors work we can then use some of the words we use to talk about length when we want to say something about time. In support of the hypothesis it can be noted that we speak of 'longer' and 'shorter' intervals of time, a 'distant' future; and we say that we have some tasks 'in front of' us, that some events are 'behind' us, and that we are looking 'forward' to doing something. Here the structure underlying the length dimension is transferred to the time dimension and we know what the words mean as expressions about time since we can identify the corresponding structure on the conceptual time dimension.

I propose that the length dimension is the more fundamental one and these expressions are thus used metaphorically for the time dimension. This may be difficult to see since these expressions about time are so idiomatic in our language that we no longer think of them as metaphors. However, their origin as metaphors can be highlighted by comparing our time expressions to those of other cultures. We need not go very far from the standard Western view of time; a particularly revealing example can be found in the ancient Greek conception of time. The Greeks thought of time as a river flowing past us. We sit in the river with our backs towards the future and see the events pass by, become distant and eventually disappear in oblivion. But we do not see what is coming. Indeed, one Greek word for 'future' (οπισφε) is the same as the word for 'behind'! In contrast, we think of time rather as a road we are travelling along in which the past is 'behind' us and we are looking 'forward' to the future. Both the ancient Greeks and we view time as a one-dimensional line (in contrast to a circular conception of time as in some cultures) – the only difference is the 'direction' of the line.

This is an example of how a study of the basic metaphors of a language can reveal the structure of the underlying conceptual dimensions. Another linguistic category that is essentially metaphorical is the class of *prepositions*. Words like 'in', 'at', 'on', 'under' etc. originate in spatial metaphors and when combined with non-locational words they create a 'spatially structured' mental representation of the expression. Herskovits (1986) presents an elaborated study of the fundamental spatial meanings of prepositions and she shows how the spatial structure is transferred in a metaphoric manner to other contexts. A sentence like "We meet *at* six o'clock" provides a further illustration to the dependence of temporal language on spatial dimensions (Herskovits (1986), p. 51). Here "six o'clock" is conceived as a point on a travel trajectory, and the locational preposition 'at' is used in exactly the same way as in "The train is at the bridge".

The theory presented here seems to go along the same lines as the one developed by Lakoff and Johnson (1980). They analyse several networks of metaphors used to talk about special topics. Among other things, they argue, in line with the description above, that the introduction of a new metaphor *creates* similarities of a new kind. These similarities are not 'objective', but, once one quality dimension has been connected to another via a metaphor, this connection may serve as a generator for new metaphors based on the same kind of similarity.

A closely related point is raised by Tourangeau and Sternberg (1982). Their 'domains-interaction' view is based on the observation that "metaphors often involve seeing in a new way not only two particular things but the domains to which they belong as well. ... Metaphors can thus involve whole systems of concepts" (p. 214). In other words, *a metaphor does not come alone* – it is not only a comparison between two single concepts, but involves an identification of the structure of two quality dimensions. Black (1979, p. 31) makes essentially the same point by the phrase "Every metaphor is the tip of a submerged model."

Tourangeau and Sternberg's analysis is obviously congenial to the present one. They even use the notion of 'dimension' when spelling out their view. Also Indurkhya (1986) interprets metaphors in terms of mappings between different domains. However, the topological structure of the domains is not exploited.

These arguments can only indicate the general direction of a systematic analysis of metaphors. Further examples and an extended analysis can be found in Gärdenfors (1992). I hope they show that an analysis of metaphors in terms of similarities of topological structures between dimensions is a promising program.

VII. PROTOTYPE THEORY

Finally, in order to say something about how slogan VI can be treated within a cognitive semantics, I want to show that describing properties as convex regions of conceptual spaces fits very well with the so called *prototype theory* of categorization developed by Rosch and her collaborators (Rosch 1975, 1978, Mervis and Rosch 1981, Lakoff 1987). The main idea of prototype theory is that within a category of objects, like those instantiating a property, certain members are judged to be more representative of the category than others. The most representative members of a category are called *prototypical* members.

Now, if a traditional definition of a property is adopted, it is very difficult to explain such prototype effects. Either an object is a member of the class assigned to a property (relative to a given possible world), or it is not, and all members of the class have equal status as category members.

Rosch's research has been aimed at showing asymmetries among category members and asymmetric structures within categories. Since the traditional definition of a property neither predicts nor explains such asymmetries, something else must be going on.

In contrast, if properties are described as convex regions of a conceptual space, prototype effects are indeed to be expected. In a convex region one can describe positions as being more or less *central*. For example, if color properties are identified with convex subsets of the color space, the central points of these regions would be the most prototypical examples of the color. In a series of experiments, Rosch has been able to demonstrate the psychological reality of such 'focal' colors.

For more complex categories like 'bird' it is perhaps more difficult to describe the underlying conceptual space. However, if something like Marr and Nishihara's (1978) analysis of shapes is adopted, we can begin to see how such a space would appear.[8] Their scheme for describing biological forms uses hierarchies of cylinder-like modelling primitives. Each cylinder is described by two coordinates (length and width). Cylinders are combined by determining the angle between the dominating cylinder and the added one (two polar coordinates) and the position of the added cylinder in relation to the dominating one (two coordinates). The details of the representation are not important in the present context, but it is worth noting that on each level of the hierarchy an object is described by a comparatively small number of coordinates based on lengths and angles. Thus the object can be identified as a hierarchially structured vector in a (higher order) conceptual space. Figure 3 provides an illustration of the hierarchial structure of their representations.

Furthermore, a 'prototypical' vector for an object category like 'bird' identifies a spatial structure that can serve as an image schema for that category. Such an image schema represents a *basic level* category (cf. Section 3), while subordinate categories like 'ostrich' are represented by subregions of the convex region associated with the prototypical object. Superordinate categories like 'animal' do not have any associated image schemas. This way of representing object categories can form a foundation for an explanation of many of the characteristics of basic level categories.

It should be noted that even if even if different members of a category are judged to be more or less prototypical, it does not follow that some of the existing objects must represent 'the prototype'. If a category is viewed as a convex region of a conceptual space, this is easily explained, since the central member of the region (if unique) is a possible individual in the sense discussed above (if all its properties are specified) but need not be among the existing members of the category. Such a prototype point in the region need not be completely described as an individual, but is normally

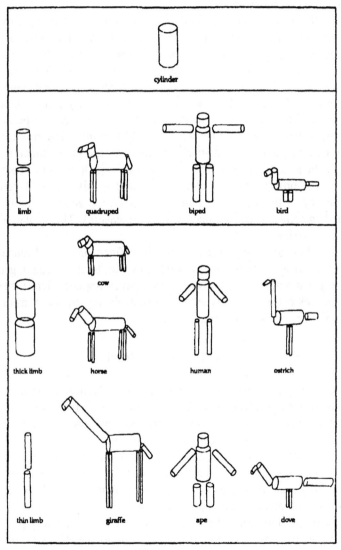

Figure 3.
Hierarchical representation of animal shapes using cylinders
as modelling primitives (from Marr (1982))

represented as a partial vector, where only the values of the dimensions that are relevant to the category have been determined. For example, the general shape of the prototypical bird would be included in the vector, but its color or age would presumably not.

It is possible to argue in the converse direction too and show that if prototype theory is adopted, then the representation of properties as convex regions is to be expected. Assume that some quality dimensions of a conceptual space are given, for example the dimensions of color space, and that we want to partition it into a number of categories, for example color categories. If we start from a set of prototypes $p_1, ..., p_n$ of the categories, for example the focal colors, then these should be the central points in the categories they represent. One way of using this information is to assume that for every point p in the space one can measure the *distance* from p to each of the p_i's. If we now stipulate that p belongs to the same category as the *closest* prototype p_i, it can be shown that this rule will generate a partitioning of the space that *consists of convex areas* (convexity is here defined in terms of the assumed distance measure). This is the so called *Voronoi tessellation*.

Thus, assuming that a metric is defined on the subspace that is subject to categorization, a set of prototypes will, by this method, generate a unique partitioning of the subspace into convex regions. Hence there is an intimate link between prototype theory and analysing properties as described by convex regions in a conceptual space.

VIII. CONCLUSION

It is becoming more and more obvious that the classical view of semantics leads to serious problems when one tries to apply it to various features of natural languages. I have presented a brief sketch of an alternative position that has become known as cognitive semantics.

As an ontological foundation for a cognitive semantics, I propose, that conceptual spaces be used as a framework for representing information. I have outlined the first steps of a cognitive semantics based on conceptual spaces. Furthermore, I have argued that the conceptual spaces are useful for understanding some semantic areas that have been particularly problematic for the classical view, namely metaphors and prototype effects in concepts*.

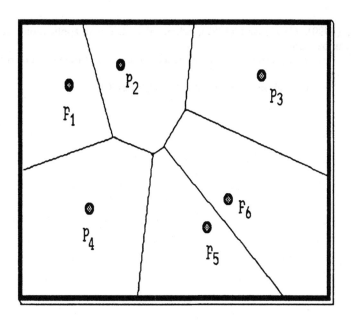

Figure 4
Voronoi tessellation of the plane into convex sets.

NOTES

* Acknowledgement. Research for this paper has been supported by the Swedish Council for Research in the Humanities and Social Sciences.

[1] Sections 2 and 4-7 in this paper are based on Gärdenfors (1991) and (1992).

[2] For further discussion of the problems of the traditional account of properties in connection with induction, cf. Gärdenfors (1990) and (1991).

[3] For a discussion of the implication sfor semantics of this translatability, cf. Jackendoff (1987).

[4] See Smith and Medin (1981) for a presentation of this and other theories of concept formation.

[5] See e.g. Rosch (1975), (1978), Mervis and Rosch (1981), Smith and Medin (1981), and Lakoff (1987) for extended discussions of the theory.

[6] For further details of the theory of conceptual spaces, cf. Gärdenfors

(1988), (1990), (1991) and (1992), where also different applications of the theory can be found.

[7] Abstract entities may be assigned values on a different set of quality dimensions.

[8] This analysis is expanded in Marr (1982), Ch. 5.

REFERENCES

Black, M.: 1979, "More about metaphor," in *Metaphor and Thought*, ed. by A. Ortony, Cambridge University Press, Cambridge, pp. 19-45.

Brugman, C.: 1981, *Story of Over*, Indiana Linguistics Club, Bloomington.

Declés, J.-P.: 1985, *Représentation des connaissances*, Actes Semiotiques - Documents, VII, 69-70, Institut National de la Langue Française, Paris.

Fauconnier, G.: 1985, *Mental Spaces*, MIT Press, Cambridge, MA.

Fodor, J. A.: 1981, *Representations*, MIT Press, Cambridge, MA.

Gärdenfors, P.: 1988, "Semantics, conceptual spaces and the dimensions of music" in *Essays on the Philosophy of Music*, ed. V. Rantala, L. Rowell, and E. Tarasti, (*Acta Philosophica Fennica*, vol. 43), Helsinki, pp. 9-27.

Gärdenfors, P.: 1990, "Induction, conceptual spaces and AI" *Philosophy of Science* 57, pp. 78-95.

Gärdenfors, P.: 1991, "Frameworks for properties: Possible worlds vs. conceptual spaces" *Language, Knowledge and Intentionality (Acta Philosophica Fennica, vol. 49)*, ed. by L. Haaparanta, M. Kusch, and I. Niiniluoto, Helsinki, pp. 383-407.

Gärdenfors, P.: 1992, "Mental representation, conceptual spaces and metaphors" to appear in *Synthese*.

Herskovits, A.: 1986, *Language and Spatial Cognition: An Interdisciplinary Study of the Prepositions in English*, Cambridge University Press, Cambridge.

Indurkhya, B.: 1986, "Constrained semantic transference: A formal theory of metaphors" *Synthese* 68, pp. 515-551.

Jackendoff, R.: 1983, *Semantics and Cognition*, MIT Press, Cambridge, MA.

Jackendoff, R.: 1987, "On Beyond Zebra: The relation of linguistic and visual information" *Cognition* 26, pp. 89-114.

Jackendoff, R.: 1990, *Semantic Structures*, MIT Press, Cambridge, MA.

Johnson-Laird, P. J.: 1983, *Mental Models*, Cambridge University Press, Cambridge.

Lakoff, G.: 1987, *Women, Fire, and Dangerous Things*, University of Chicago Press: Chicago, IL.

Lakoff, G. and Johnson, M.: 1980, *Metaphors We Live By*, University of Chicago Press, Chicago, IL.

Langacker, R. W.: 1986) *Foundations of Cognitive Grammar, Vol. 1)*, Stanford University Press, Stanford, CA.

Marr, D. and Nishihara, H. K.: 1978, "Representation and recognition of the spatial organization of three-dimensional shapes" *Proceedings of the Royal Society in London, B 200*, pp. 269-294.

Marr, D.: 1982, *Vision*, Freeman, San Francisco.

Mervis, C. and Rosch, E.: 1981, "Categorization of natural objects" *Annual Review of Psychology* 32, pp. 89-115.

Montague, R.: 1974, *Formal Philosophy*, Ed. by R. H. Thomason, Yale University Press, New Haven.

Petitot-Cocorda, J.: 1985, *Morphogenèse du Sens I*, Presses Universitaires de France, Paris.

Putnam, H.: 1981, *Reason, Truth, and History*, Cambridge University Press: Cambridge.

Rosch, E.: 1975, "Cognitive representations of semantic categories" *Journal of Experimental Psychology: General* 104, pp. 192-233.

Rosch, E.: 1978, "Prototype classification and logical classification: The two systems" *New Trends in Cognitive Representation: Challenges to Piaget's Theory*, ed. E. Scholnik, Lawrence Erlbaum Associates: Hillsdale, NJ, pp. 73-86.

Smith, E. och Medin, D. L.: 1981, *Categories and Concepts*, Harvard University Press, Cambridge, MA.

Stalnaker, R.: 1981, "Antiessentialism" *Midwest Studies of Philosophy 4*, pp. 343-355.

Sweetser, E.: 1990, *From Etymology to Pragmatics*, Cambridge University Press, Cambridge.

Talmy, L.: 1988, "Force dynamics in language and cognition" *Cognitive Science* 12, pp. 49-100.

Tourangeau, R. and Sternberg, R. J.: 1982, "Understanding and appreciating metaphors" *Cognition* 11, pp. 203-244.

Cognitive Science
Department of Philosophy
Lund University
S-223 50 Lund, Sweden
E-mail: Peter.Gardenfors@fil.lu.se

KEITH STENNING

THE COGNITIVE IMPACT OF DIAGRAMS*

I INTRODUCTION

The purpose of this paper is to give an introduction to a general theory of
graphical communication, and to illustrate its approach through an example
of a system of elementary logic diagrams developed from Euler's Circles. By
a theory of graphical communication I mean a theory which can predict and
explain the cognitive consequences of formulating a message in graphical
media as opposed to expressing the same information in language. In fact,
most graphical communication includes the use of language, and so the
theory will have to deal with combinations of language and graphics. Such a
theory is intended to form the basis of practices of design of optimal
messages and therefore has substantial practical importance in, for example,
the design of publications, teaching materials and computer interfaces.

The theory and the example analysis of Euler's Circles has been
presented in considerably more detail than there is room for here in Stenning
& Oberlander (in press) and Stenning & Oberlander (in submission). The
purpose here is to describe the theory for a wider audience which may be
interested in the general approach.

In the early eighteenth century, Bishop Berkeley discussed the issue
of interpretation of visual information in an essay entitled 'A new Theory of
Vision' and elsewhere in his philosophical works. The context of Berkeley's
discussion was Locke's 'picture theory of meaning' -the theory that the
meaning of a word was a mental image of the thing denoted by the word
which was brought to mind by hearing the sound of the word. Berkeley was
the first to point out a fundamental problem with such theories which derives
from what we call the 'specificity of graphics'. Berkeley's objection to
Locke's picture theory was that pictures were too overdetermined to achieve
the sort of abstraction involved in linguistic meaning. A picture of a triangle
has to have determinate angles and ratios of sides. It has to be either irregular,
isosceles or equilateral. Yet, the word *triangle* abstracts over these properties
of particular triangles. (In case distance in history is taken to indicate
irrelevance to cognitive science, it is worth pointing out that Berkeley's
objections to Locke are not so very distant from Pylyshyn's objections to

181

A. Clark et al. (eds.), Philosophy and Cognitive Science, 181–196.
© 1996 Kluwer Academic Publishers.

psychologists' theories of imagery lodged in his 1973 paper 'What the mind's eye tells the mind's brain'). This fundamental feature -the specificity of graphical representations- we believe lies at the root of their cognitive properties. Graphical representation systems enforce the representation of certain classes of information. Languages, at least fully expressive languages such as English, always allow the avoidance of expression of any class of information.

Berkeley's observation was coupled with another balancing observation which stemmed from his interest in mathematical, and particularly geometric, proof. He observed that geometry proofs are generally accompanied by diagrams representing, for example, particular triangles, and this gives rise to a puzzle about how such proofs can arrive at general conclusions. Here, the conventions of interpretation play a critical role. In learning geometry, one has to learn that only some properties of diagrams can be appealed to in proofs -or that the properties appealed to have an impact on the generality of the conclusions that can be drawn. If one appeals to the particular value of an angle in constructing ones proof, it will hold for that value, but not necessarily for any others. If one wants a proof general to all triangles, then appeal must only be made to properties of the diagram which are general to all triangles. These rules of valid proof are backed by conventions about which sorts of diagrams are used to illustrate them -in general, if a proof is supposed to be general to all triangles, an irregular triangle is chosen rather than an equilateral one.

These observations raise interesting cognitive questions about just what good the diagram is. The observation that even the simplest proof is intractable without the guidance of a diagram is balanced by the fact that only quite abstract properties of the diagram can apparently be appealed to in proofs. So the observation that graphical representations enforce representation of information is balanced by the observation that actual systems of graphical representation usually reintroduce abstraction through conventions of construction and interpretation. It appears to be the tension between these two observations that has prevented anyone from building a theory of graphical communication on them. There is a strong flavour of 'the lord giveth and the lord taketh away' about these twin intuitions. *If* the abstractions which cannot be expressed directly by diagrams can *all* be reintroduced by conventions of interpretation, then the graphical systems remain equivalent to linguistic ones at this level of analysis. This offers no basis for understanding differences between graphical and linguistic communication and reasoning.

We believe that this is a mistaken pessimism about both of these central intuitions. This pessimism relies on a sort of conjuring trick in which there is very definitely a rabbit deposited in the hat during what appear to be equal and opposite movements. The abstractions that can be reintroduced by conventions of interpretation are not *all* of the abstractions which have to be given up in resorting to diagrams. The resulting representational systems are not completely concrete but they are still severely curtailed in their expressive power, and it is this curtailment which gives them their advantages (and their disadvantages too). Clearly what is required to turn these twin intuitions into an honest theory is a specification of what abstractions cannot be expressed by graphical systems of representation; and what abstractions can be reintroduced by conventions of interpretation. Such a specification gives a characterisation of a representation system in logical terms, and we can then relate that system to the complexity of inferences involved in the system's use through what is known of the computational complexity of logics. Logics which are less expressive are more inferentially tractable whether they are implemented in machines or in minds.

In Stenning & Oberlander (in submission) we have proposed a logical framework for analysing these phenomena. Here I will proceed rather by example. Euler's Circles are a system designed for teaching syllogistic logic by Euler in the 18th century. Venn, Carroll and Peirce each developed graphical systems for logic. Venn's system is closely related to Euler's and is now better known amongst logicians. We choose Euler's system for analysis because of its psychological characteristics which are quite different from the Venn system. In fact, several psychological models have been based on Euler's Circles (e.g. Erickson 1974, Newell 1981, Guyote & Sternberg 1981) and have made claims that the mental representations people employ in doing syllogistic reasoning 'in their heads' are just the ones which Euler described. Other psychologists, notably Johnson-Laird 1983, have claimed that simple arguments about the combinatorics of Euler's system are sufficient to rule out these theories and have proposed other systems which they claim are distinct.

Euler's Circles are therefore fertile ground for a cognitive theory of the distinctive properties of graphical representation. Syllogistic logic is a paradigm example of a verbal reasoning task. Here is a system of graphical reasoning for which there are two centuries of didactic evidence that it aids learning and performance. There is even a theoretical literature claiming to produce behavioural evidence for Euler's Circles' status as mental representations, and the invention of an alternative linguistically based system based on the claimed intractability of the graphical system.

Analysis of Euler's Circles using the approach to graphical representations sketched above reveals that in fact both the psychological models which use ECs and the arguments that they are intractable are based on a misinterpretation of the system, and that this misinterpretation is directly related to the primary distinction which we make between primitive concretely interpreted graphics on the one hand, and the limited abstractions which interpretation conventions permit on the other. The psychological models assume that ECs are primitively interpreted and the combinatoric explosion which is supposed to rule them out is a direct result of this misinterpretation. In fact, the conventions of interpretation employed by logicians in using ECs and analysed here allow them to express just the abstractions required for syllogistic logic but this expressive power is not sufficient for the whole of monadic predicate calculus, or even the disjunctive syllogism.

This case study therefore gives some evidence that the framework provides useful insights into real graphical systems. If these results can be extended, they throw light on a much more general cognitive phenomenon, that of analogical reasoning. Graphical systems of representation in general, and ECs in particular are cases of analogy. Such spatial analogies are endemic in human reasoning. Our theory proposes that the computational value of analogies is that they provide a more restricted, less expressive 'language' which is more inferentially tractable so long as the problem at hand can be cast within the available resources. This proposal brings the study of human thinking into an interesting correspondence with an important issue in computer science.

Computer scientists develop highly expressive specification languages for communicating about the desirable properties of systems to be implemented. But these languages are intractable for the conduct of the reasoning which the system they specify has to perform. For this, the system has to be implemented in some less expressive but more tractable language which can be executed. Considerable attention is being focussed on the problems of proving that an executable program actually implements a given specification.

This relation between specification and execution languages is analogous to the relation in human thought between natural languages and specialised graphical notations such as Euler's Circles and other spatial analogies. Natural languages are powerfully expressive but also highly intractable for inference. Finding an analogy which allows a problem stated in natural language to be recast into graphical representations is like implementing a piece of specification -it allows reasoning to be carried out much more efficiently. This is not to suggest that graphical analogies are the exclusive medium of reasoning- merely that the media of thought are likely to be much less expressive systems of representation than natural languages.

Our example domain of logic diagrams provides an unusual illustration of how the relation between specification and execution languages relates to human thinking. Most of the literature on analogy understandably explores rich analogies and their role in learning (e.g. Gentner & Stevens 1981). In these analogies (say that between hydraulic and electrical flow) at least one term of the analogy is ill-understood. What is particularly instructive about ECs is that they are an analogy for which the users have a perfect grasp of both terms from the outset -undergraduate students do not need teaching about spatial inclusion or set-membership in order to apply the system, just as they do not need teaching about the meaning of 'all' and 'some'. What they do need, and what the system facilitates, is a tractable method of reasoning about a particular fragment of language called the syllogism. Studying this fragment through graphical algorithms reveals important logical properties which explain why it can be implemented graphically, and why other small fragments cannot. Understanding of graphical reasoning in elementary logic suggests the hypothesis that the specificity of graphics is what limits its logical expressiveness and that in turn is what aids reasoning. This is an interesting hypothesis about the role of analogy more generally in thinking.

The plan of the paper is as follows. The next section gives a brief overview of the Euler's Circle system. Section 3 relates the Euler's Circles system and its graphical algorithm to the general theory of graphics outlined above. We examine how the circles force the representation of certain information and how conventions of interpretation allow abstraction and prevent combinatorial explosion.

II EULER'S CIRCLES

The possible relations between two sets which are relevant to syllogistic logic can be portrayed by the five topological arrangements of circles shown in Figure 1

These relations map in a many-to-one fashion onto the four possible rclations expressed by the quantifiers of the syllogism. It is this mapping which gets a primitive interpretation of Euler's Circles into trouble as a model for syllogistic logic. If the diagrams are interpreted as representing the existence of each type of entity that corresponds to a region in the diagram, and each statement is represented by all possible diagrams which model it, there are as many as four diagrams for each linguistic premiss ('Some A are B' maps onto four diagrams). Erickson's model has to posit stochastic choice in order to avoid this explosion of diagrams -Johnson-Laird points out that this means the model cannot give an adequate *competence* model of syllogistic reasoning because it only ever gets right answers by chance (even if it gets them as often as subjects do). Johnson-Laird concludes that no model can be based on Euler's Circles and goes on to invent Mental Models, a system which packs several models into each representation.

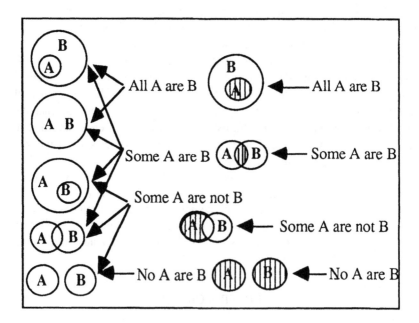

Figure 1. Mapping diagrams to sentences under a primitive and a
sophisticated interpretation.

However, this is not how logic teachers interpret Euler's Circles. They first teach that the strategy for diagram choice is to pick the *maximal* model (the one with the most types consistent with the premiss, shown on the right in Figure 1). This strategy is repeated when the two premiss diagrams are combined -the B circles are 'registered' (they are, after all, the same circle) and the A and C circles are arranged so that the maximal number of regions are represented, consistent with each individual premiss[1]. Examples of the resulting *registration* diagrams for syllogisms both with and without valid conclusions appear in Figure 2.

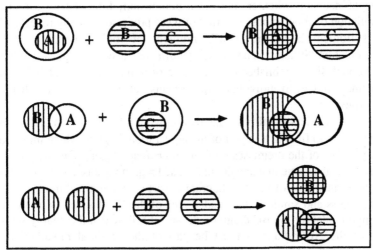

Figure 2: Premiss and registration diagrams for example syllogisms:
All A are B. No B are C; Some B are not C.
All C are B; and No A are B. No B are C.

This strategy of maximal model selection goes along with an attendant shift in interpretation (remember Berkeley's triangles). Instead of interpreting a region as representing the existence of a type of individual a region is now interpreted as representing the *possibility* of the existence of a type of individual. Thus, in an indirect way, the diagram represents all *consistent* interpretations of the premisses. This shift in interpretation is often supported by a notation of shading the *minimal* models of the statement -the models containing the least number of types consistent with the premisses. Thus the shaded diagram characterising 'All A are B' represents the fact that a model containing just ABs is minimal, and a model containing ABs and A¬Bs is also possible, and is in fact maximal. Thus one diagram in a sense represents two models of the sort represented by the primitively interpreted diagram.

This economy of representation restores the one-to-one mapping between diagrams and premisses. The same strategy of registering pairs of premiss diagrams to derive the maximal conjoint diagram also means that the construction of a diagram to represent any syllogism is entirely deterministic. It remains to explain how to make inferences from these diagrams about valid conclusions. This process is usefully broken down into two stages---first a stage of decision whether there is a conclusion, followed by linguistic formulation of conclusions. The first of these processes actually reveals why the syllogism is a 'graphical' fragment of logic amenable to EC representation. Syllogisms are 'case identifiable' -every syllogism that has a valid conclusion establishes the existence of a maximal type of individual (maximal applied to individual types means defined in terms of all three properties)[2].

Case identifiability means that drawing inferences from syllogisms comes down to finding a region of the registration diagram which must exist in any model of the premisses (call this a *critical* region). One may at first be tempted to assume that any double-shaded region must be a critical region---if the shaded area of the AB diagram is part of the minimal model for the first premiss and similarly the shaded area of the BC diagram is part of the minimal model for the BC diagram, doesn't it follow that the doubly shaded area of the ABC diagram must be part of the minimal model of both premisses? The answer is 'no' but in fact subjects' errors appear to be influenced by this conjecture -the hardest syllogism that does have a valid conclusion is the only one which has no doubly shaded region. Subjects often believe that this syllogism has no conclusion. The hardest syllogism with no valid conclusion (which subjects often draw erroneous conclusion from) is one which does have a doubly shaded area.

So if this is not an adequate algorithm for finding critical areas, how are we to do it? It turns out that critical regions are ones which are shaded in the parent premisses and remain un-bisected by the third circle during registration (I leave as an exercise the question of why the logical model theory corresponds to this topological property -see Stenning & Oberlander (in submission)). If there is no critical region, then there is no valid conclusion.

At this point we will look at some behavioural evidence that this algorithm is related to human performance without external diagrams. We do this to motivate the study of external graphical systems as guides to internal processing. Sometimes critical regions come from the first premiss, sometimes from the second and sometimes from both. Yule & Stenning 1992 show that this variation can be used to generalise a phenomenon which Johnson-Laird & Steedman 1978 reported -the figural effect. The most insightful statement of the figural effect is that subjects prefer conclusions which conserve the grammatical category of terms in the premisses in the conclusion. Thus if A is a subject term and C is a predicate term, then AC conclusions are preferred, and if the categories are reversed, CA conclusions are preferred. This regularity has been observed many times in the literature, but this statement of the regularity obscures that it is the logical rather than the grammatical properties of terms which drives the phenomenon. The graphical approach, with its focus on critical regions, reveals that the fundamental regularity is that terms from the premiss which establishes the critical region are preferred in subject position. This graphical formulation can, of course, be translated into a logical formulation.

The grammatical formulation of the figural effect can only be applied to the 'diagonal' figures (AB, BC and BA, CB) because, in the 'columnar' figures, both terms have the same grammatical category. However, the property of being, or not being, a term in the premiss establishing the critical region can be applied to terms in all four figures. Yule & Stenning show that terms with this property are preferred in subject position thus subsuming the 'figural' effect under the 'critical individual' effect. This reconceptualisation relates the effect to the important process of general text comprehension by which individuals are introduced into the interpretation of the text, rather than to a parochial syntactic property of syllogisms.

Further evidence that subjects employ algorithms organised around critical regions (or their corresponding types of individual) is provided by Yule & Stenning's task. Instead of asking subjects to draw conventional syllogistic conclusions from syllogistic premisses, the new task asked for descriptions of critical individuals[3]. Subjects are in general better at this task than the logically more complex conventional one and were quite good at finding the valid conclusions which are inexpressible in the conventional notation (for the syllogisms mentioned above which determine critical regions but do not have conventional conclusions). Even more suggestive is the fact that the order of the three predicates in subjects' descriptions of critical individuals accord closely with a simple model based on the graphical algorithm which contrasts radically with Johnson-Laird's (1983) predictions based on mental models.

Having found a critical region of a registration diagram, formulation of conclusions is straightforward. Existential conclusions can be derived by prefixing a quantifier to the description of the individual corresponding to the critical region, dropping out the B term, and ensuring that the subject term is not negative by switching the order of the terms if necessary. For example, the type-description ¬C B A gives rise to the conclusion *Some A are not C* . Universal conclusions are only warranted if the critical region is circular and labelled by a non-middle term. This labelling term then becomes the subject of the universal conclusion. For example, from a circular region labelled C , corresponding to B C A , we can conclude *All C are A* .

This completes our graphical algorithm for reasoning with Euler's Circles. It is summarised for convenience in Table 1.

1.	Form characteristic diagram for each premiss,
2.	Register B circles of the characteristic diagrams of the premisses and arrange A an C circles with most types consistent with the premisses.
3.	If no shaded region from a component premiss remains non-intersected, then exit with No Valid Conclusion response. If there is one, then it is the *critical* region.
4.	If such a region does exist but both premisses are negative, then exit with a No Conventional Valid Conclusion response. (If task permits, conclude that *Some non-As are not Cs*).
5.	Formulate conclusion:

(a)   Take the description of the individual type represented by the critical region of the diagram (e.g. A ¬BC)

(b)   Eliminate the B term from this description

(c)   Existentially quantify the remaining description for an existential response

(d)   Is the critical region circular and labelled by an end-term?

       i.   If so, it is the subject term of a universal conclusion

       ii.   If not, there is no universal conclusion

III SPECIFICITY AND ABSTRACTION IN THE EULER'S CIRCLES SYSTEM

How does this example relate to the general theory of graphical representations sketched in the Introduction? What abstractions do the graphical representations prevent us expressing, and what conventions of interpretation reinstate which of these abstractions?

The most fundamental specificity of the graphical system is its enforcement of representation of maximal types of individual and this flows directly from a simple topological property of planes-any three circles in a plane define every point in the plane with regard to whether they are inside or outside each circle. So every point in a plane containing three circles is defined to correspond to a maximal type of individual. Individual types that are less than maximally specified cannot be represented in a completed registration diagram. It is this fact, together with the assumption that the basic binding mechanism of human working memory holds total mappings of attributes to individuals, that is the centrepiece or our explanation of the efficacy of graphical algorithms in this domain. Formulating linguistic conclusions for output requires abstracting away from critical individuals by removal of the middle term, but this is at the stage where conclusions are being read from the agglomerative representation. It is this fundamental specificity which prevents a direct implementation of a natural deduction-type of proof theory for the syllogism in a graphical system.

Examining the graphical system in greater detail reveals other specificities. If we look in more detail at the actual processes of registration of diagrams, analysing them as real mechanical motions rather than unanalysed logical operations, we find important computational properties of the graphical medium in time -what is often called 'animation'. In registering two premiss diagrams, as we slide the two B circles into correspondence and arrange the A and C circles into the maximal model, each time step is a static diagram and the specificities arise in the time dimension in just the same way as they do in spatial relations. Just as any two points in a diagram are determined in their spatial relations, so any two steps are properly ordered in time.

This property of the medium can be exploited because of a logical property of the interpretations -movements of circles either add or eliminate regions from the diagrams one at a time[4]. Logically 'relevant' alternative models are always neighbours in the space of models represented by a registration diagram.

Beyond enforcing the maximality of all individual types represented in registration diagrams, there are some logical models of the syllogism which cannot be represented by any diagram as long as we stick to circles. Of the 2^7 models, only about half have Euler's Circles (see Stenning & Oberlander (in press) for more detail). So the diagrams are quite inexpressive about configurations of maximal individuals as well as about the degree of individuals' specification.

What conventions allow back what abstractions? The most important convention is the interpretation of regions as denoting *possible* individuals rather than actual individuals. This convention, coupled with the strategic choice of *maximal* models to represent premises and in the registration process, is what ensures that reasoners only need construct a single registration diagram. Without this interpretation and strategy the graphical representations suffer combinatorial explosions which are direct results of their specificity -if only total models can be represented, and all possible models have to be considered, there is a large space of models to represent.

The shading notation aids in the identification of critical regions. Euler does not appear to have used any equivalent notation -it is not really needed if the teacher teaches the strategy of remembering which regions give rise to necessary types of individual. Certainly the convention has been invented many times. We have observed subjects in the process of invention and have been told by students that it was used in some 'new maths' approaches to teaching in elementary school.

If these are the specificities which Euler's Circles impose which deny the expression of abstractions, and these are the conventions which reintroduce some abstractions, what logical fragment remains expressible by the Euler's Circles system? The fragment is greater than the classical syllogism-for example, it can express the 'unconventional' conclusions studied by Yule & Stenning and can handle arguments with more than two premises. Yet it cannot contain the whole monadic predicate calculus. As we have seen it cannot distinguish each of the models of the syllogism and certainly cannot express disjunctive syllogisms (because they are not case-identifiable). Case identifiability appears to be the defining property of the fragment. It is interesting that Euler's Circles *can* , contrary to Johnson-Laird, Byrne and Tabossi 1989, capture the multiple quantifier relational arguments which they study, and this is because those arguments easily reduce to case-identifiable monadic (in fact syllogistic) arguments. To our knowledge the logical property of case-identifiability has not been studied by logicians. It is the restriction on the representation of bindings between properties to those of maximal individuals which leads to the definition of the case-identifiable fragment as the implementable one.

It is perhaps worth concluding with the comment that graphical methods in general are not as limited as Euler's Circles. For example, Sun-Joo Shin (1992) has shown that an extension of Venn Diagrams can capture the whole monadic predicate calculus. However, it is psychologically interesting to note that the system has to introduce much more complex notational additions and conventions for their interpretation, and in doing so intuitively gives up many of the cognitive benefits of the graphical medium.

The Euler's Circles example serves as a useful illustration of the general framework for treating graphical systems. Stenning & Oberlander (in submission) gives a more general logical framework for developing the theory.

IV DISCUSSION

This paper has been almost exclusively concerned with the properties of an external graphical system (a pencil and paper technique) which faithfully implements a fragment of logic. Its purpose has been to illustrate a general approach to understanding the computational and cognitive properties of graphical representations -an approach which rests on two very old intuitions. I hope it has succeeded in persuading the reader that detail can be hung on these intuitions to yield a well-specified computational theory of graphical communication for at least this example domain.

The paper has made some mention of human performance at the task of solving syllogisms when paper and pencil are not available. One long term interest is in internal representations which function in this situation but here we have only made a few gestures toward behavioural phenomena which appear to be easy to relate to this style of graphical algorithm. More serious study requires definition of a performance model with error generating processes (mistaken strategies, memory overloads etc. etc.) which will be reported elsewhere. We will finish with a few comments about the relation between internal and external representations.

Understanding a felicitous external system of representation should certainly help with the problem of understanding internal representations which are involved when the same task is performed 'mentally'. However, the relation between external and internal representations cannot be taken to be simple isomorphism. In particular, we would not assume that phenomenology will be a reliable guide to whether or not graphical representations (in our sense) are involved. Our whole approach to graphics is through general computational properties and seeks to capture processing characteristics by specifying graphical representation systems as 'languages' of restricted expressiveness. It goes without saying that these logics can then be implemented in a variety of ways, some of which are definitely not 'picture-like'.

What we do know about the implementation of algorithms like our graphical method suggests that the nature of variable binding in human working memory may be what lies behind the adoption of this strategy of reasoning by people but much remains to be done to show whether and how these algorithms are implemented mentally (see Stenning & Oberlander (in press) for a detailed discussion about implementation). The 'animation' characteristics of Euler's Circles are very suggestive that humans may be able to exploit mental mechanisms which were developed for understanding and controlling movements of shapes in our mental use of Euler's Circles. However, our understanding of spatial working memory is not yet at a sufficiently advanced state to be able to relate what is known to this particular diagrammatic system.

We believe graphical systems of representation which have proved their usefulness in real-world tasks deserve this sort of careful computational analysis. Apart from insight into graphical communication, such analyses will allow us to construct theories of how diagrams are mentally represented in human working memory, an important issue for theories of imagery. A cognitive theory of graphics would be an important step towards understanding human inference and toward designing better information displays.

NOTES

*This work was partially supported by grant #9018050 from the joint Councils Inititative in Cognitive Science an HCI. the partial support of the Economic Research Council UK (ESRC) is also gratefully acknowledged. the work was part of the research program of the ESRC funded Human Communication Research Centre (HCRC).

[1]   This strategy assumes that subjects can assess the relation between pairs of circles and statements of the logic for consistency. We assume that this *is* easy for subjects and that it is the *combination* of premises and the inferences

about the necessary relations between A and C which present the problems

^{2} In fact the biconditional is very nearly true but instructively false---there is only a small class of syllogisms which establish maximal individuals but do not have valid conclusions expressible in Aristotles' system (see Stenning&Oberlander (in press) for a discussion of these inexpressible conclusions and (see Stenning&Oaksford (in press) for further discussion of case-identifiability).

^{3} The instructions asked subjects to describe any individuals defined in terms of all three predicates or their negations which must exist if the premises are true.

^{4} Strictly speaking it is possible to add or eliminate more than one region at a time, but it is never necessary to do so.

Human Communication Research Centre
Edinburgh University

REFERENCES

Erickson, J. R.: 1974, A set analysis theory of behaviour in formal syllotistic reasoning tasks in Solso, R. (ed.) *Loyola Symposium on Cognition*, Volume 2. Hillsdale, N.J.: Lawrence Erlbaum Associates.

Gentner, D. and Stevens, A. L. (eds.): 1981, *Mental Models* ; Hillsdale, N.J.: Lawrence Erlbaum Associates.

Guyote, M. J. and Sternberg, R. J.: 1981, A transitive-chain theory of syllogistic reasoning, *Cognitive Psychology*, 13 , 461-525.

Johnson-Laird, P. N. and Steedman, M. J.: 1978, The psychology of syllogisms, *Cognitive Psychology*, 10, 64-99.

Johnson-Laird, P. N.: 1983, *Mental Models*, Cambridge: Cambridge University Press.

Johnson-Laird, P. J., Byrne, R. M. J. & Tabossi, P.: 1989, Reasoning by model: the case of multiple quantification, *Psych Rev*, 96(4), 658-673.

Newell, A.: 1981, Reasoning, Problem Solving and Decision Processes: The Problem Space as a Fundamental Category, in Nickerson, R. (ed.) *Attention and Performance* , Volume 8. Hillsdale, New Jersey: Erlbaum.

Pylyshyn, Z. W.: 1973, What the Mind's Eye Tells the Mind's Brain: A Critique of Mental Imagery, *Psychological Bulletin*, 80, 1-24.

Shin, S.: 1991, A Situation-Theoretic Account of Valid Reasoning with Venn Diagrams, in Barwise, J., Gawron, J. M., Plotkin, G. and Tutiya, S. (eds.) *Situation Theory and Its Applications*, Volume 2. Chicago: Chicago University Press.

Stenning, K. and Oberlander, J.: in submission, A cognitive theory of graphical and linguistic reasoning: logic and implementation, Research Paper No. 20, Human Communication Research Centre, Edinburgh University, Edinburgh, 1991.

Stenning, K. and Oberlander, J.: (In Press), Spatial containment and set membership: a case study of analogy at work, in *Analogical Connections.*, LEA. edited by Barnden, J. and Holyoak, K.

Yule, P. & Stenning, K.: 1992, The figural effect and a graphical algorithm for syllogistic reasoning, *Proceedings of Fourteenth Annual Conference of the Cognitive Science Society*, Bloomington, Indiana.

KURT KONOLIGE

WHAT'S HAPPENING?
ELEMENTS OF COMMONSENSE CAUSATION

ABSTRACT

Knowledge of causation is an important part of commonsense reasoning.
We use cause-and-effect analysis to understand everything from why we
caught the flu to how to make a video recorder save our favorite TV show.
Our facility can be characterized by a combination of two main capabilities:
the ability to predict the outcome of a set of causative events; and the ability
to explain given facts by postulating a set of determining causes.

Causal reasoning is most effective when combined with another
type of commonsense reasoning, the assumption of normal conditions under
which the reasoning is valid. While the existence of these assumptions has
been recognized in the philosophical literature on causation, relatively little
attention has been paid to the practical aspects of reasoning involving
causation and normal conditions: for example, what happens when some
causes affect the conditions on which other causes depend?

The field of Artificial Intelligence (AI) has evolved several
techniques for formally representing and reasoning about conditions that we
normally assume by default. In this paper we present an application of these
methods in formulating a theory that integrates causal and default reasoning.
The main structure of the theory is a default causal net representing the causal
connections among propositions in the domain. The formal theory has been
implemented using computational techniques available in AI; here we are
mainly concerned with the structure of the theory, and its relation to other
work in AI.

I. INTRODUCTION

Consider the following mundane example of commonsense reasoning about
causation (Figure 1).

A. Clark et al. (eds.), Philosophy and Cognitive Science, 197–220.

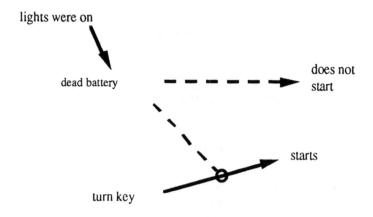

Figure 1.   Starting the car

The solid arrows represent "normal" causal connections among the propositions. Turning the key will normally cause the car to start; if the lights were on overnight, there normally will be a dead battery. There are also other kinds of information present: a dead battery means that the car will not start, and it blocks the causal relation between turning the key and starting the car. This information is represented by dashed arrows.

Now suppose we know that the lights were on overnight, and we turn the key. What conclusions should we draw? One the one hand, we can argue that the lights were on, so the battery should be dead, and so turning the key will not start the car. This is the natural conclusion to draw; but there is another one we might argue for. Suppose we start by assuming that turning the key actually will start the car; then it can't ve the case that there is a dead battery, and so perhaps leaving the lights on did not affect the battery in the normal way. Both these arguments violate one normal condition: in the first argument, the condition that turning the key normally starts the car; in the second, that leaving the lights on drains the battery.

Intuitively, we accept the first argument because, although the normal condition for starting the car is violated, there is an explanation or excuse for the violation: the battery is dead. To show this, we have drawn a dotted line in Figure 1 connecting the proposition dead battery to the causal relation between turning the key and starting the car. But there is no such excuse for concluding that the lights being on did not drain the battery. We have to invent a plausible account of how this might happen, which makes it a less persuasive argument than its competitor. In the absence of additional information, we conclude that the car will not start.

Suppose we learn that, after turning the key, the car did indeed start. Now we can no longer accept the first argument, because it leads to a conclusion we know to be false. The only other explanation of what occurred is the second argument: something must have prevented the lights being on from draining the battery.

This example illustrates some key principles of reasoning in causal domains.

1. In domains where we have incomplete information, causal reasons are subject to default assumptions for their application.

2. Defaults that lead to conflicting conclusions occur frequently, and the correct default can often be inferred from causal precedence among the defaults.

3. There may be different causal explanations for observed events; the most natural ones are those that have the fewest unexplained violations of defaults.

In the sequel we present a theory of causal and default reasoning that is based on these three principles. It is important to note that the purpose is *not* to develop a theory of causation itself by reducing it to other, more primitive concepts. This is the goal of some philosophical theories of causation, e.g., Suppes [1970] defines causation in terms of conditional probabilities of events, or Lewis [1973] in terms of counterfactual statements about possible worlds. Rather, we assume causation is a primitive relation among events, and use it to structure arguments about what defaults should apply in a given situation, and what conclusions we should accept. We call any theory that unifies causal and default reasoning a *causal default theory*. In the next two sections we develop the theory in more detail, introducing the elements of normal conditions, causal relations, and correlations.

There are two tasks that a causal default theory should address: predication and explanation. Prediction is the process of deriving the course of events from initial conditions. Prediction is useful in many ways, for example, in planning one's actions. What happens if I don't pay my telephone bill on time? Knowing the consequences of this action can help decide whether to perform it or not. Another way prediction is used is to set up expectations in testing. An electronics engineer may apply an input to a circuit, expecting it to generate a certain output if it is working correctly. In the fourth section we introduce the concept of prediction into the causal default theory, and show how the process of simulation from initial conditions can lead to correct adjudication of conflicting defaults.

The second task is explanation: from observed effects, infer what could have caused that effect. Typical here are applications such as plan recognition and diagnosis of complex systems. In plan recognition, one tries to infer the intentions of someone through observation of her actions: *Why did the train conductor ask if I had a passport?* Understanding the relation of actions to intentions is important in any cooperative task, and especially in communication [Cohen et. al. 1990]. Diagnosis is a similar kind of task, except that one is trying to figure out possible explanations for a system not behaving as expected: *Why does the copier always jam when I put in transparency paper?* Finding the answer to this question can help in fixing the problem. The fifth section is devoted to developing the concept of a causal explanation within the context of causal default theories. Finally, we discuss this work in relation to other research in AI, especially that dealing with defaults and theories of action.

II. CAUSATION AND CORRELATION

Formally, we understand causation to be a primitive relation among propositions. By "primitive" we mean that the causation relation is an unanalyzable part of the default causal theory we are developing: it is not derived from any other concepts. This approach leaves unanswered questions about how to identify causation in a given domain, the relation of causation to time, and various other difficulties about the nature and properties of causation. We will have very little more to say about these important questions, other than giving the most basic (and hopefully noncontroversial) properties. These properties suffice to develop the basic outlines of the default causal theory; further investigations will have to confront some of the other mentioned problems.

The causation relation is composed from a set of "direct" causes among propositions; we call this the one-step causation relation. Let us use the following shorthand for the propositions of Figure 1:

L	*lights switched on*
D	*dead battery*
B	*live battery*
K	*turn key*
S	*car starts*
N	*car doesn't start*

The one-step causes are:

$$L \longrightarrow D$$
$$B, K \longrightarrow S$$

There are several points to note here. First, there is a deliberate sloppiness about stating propositions. Some of them, such as "turn key," have the form of an action or event, or even a completed action ("lights switched on"). Others are simple statements of properties, e.g., "dead battery." We are trying to be as noncommittal as possible about the ontology of events and propositions, whether states of the world can be allowed as causes, how to specify the time of events, and so on. Any consistent defensible set of choices will do.

The second point is that one-step causation specifies *all* and *only* the propositions governing an effect. Turning the key only starts the car if the battery is not dead, so we have to add this condition. Of course, in any real-world situation there will be an inordinate number of such conditions, so any default causal theory will be relative to a set of background assumptions that do not enter into the theory. The choice of these assumptions is conventional; we have chosen to include key-turning and the state of the battery in the current example.

It is important that only the relevant propositions participate in one-step causation. If we add an irrelevant proposition to the antecedent of the relation, the relation would still be useful in the sense that conjunction of the antecedents produces the desired effect, but it would be misleading in implying that all the antecedents were necessary. In producing explanations (Section 4), the relevance of the antecedents is required to ensure that explanations are correct.

The causation relation is formed by taking the transitive closure of a one-step causation relation. For the above example, there are no connecting causes, and so the transitive closure is just the set of one-step causes

$$L ==> D$$
$$B, K ==> S$$

The symbol "$==>$" is used for the causation relation. The causation relation specifies both direct and indirect causes of propositions. It is the main ingredient in our formal account of "A causes B," but we need one further concept, our ability to control the outcome of a situation. We can do this by defining a set of *primitive causes* that serve as a grounding for the relation. In the example, take $C = \{L, K\}$ to be the primitive causes. The role of primitive causes is to define the propositions over which, in some sense, we can exercise direct control. The point at which we choose to define primitive causes is partly a matter of convention. Often bodily movements are taken to be the ultimate primitive causes, but this viewpoint is unnecessarily restrictive. Any well-defined event or condition that we can reliably bring about will suffice for a primitive cause, as long as the purpose of producing explanations is to give a set of conditions that account for the observed facts, and over which we have control.

If A is a set of primitive causes, and for some subset A' of A we have $A' ==> c$, then we say that A causes c, and write $A \vdash c$. One way to understand this relation is as a provability relation, with which it shares many properties. The provability relation is composed from individual inference steps combined into a tree; in the same way, the causation relation is specified by combining one-step causation into a tree structure. A proof $A \vdash c$ holds when the set A contains just axioms, and there is a subset A' for which the provability relation to c holds. Similarly, $A \vdash c$ holds when A is a set of primitive causes, and the causation relation holds between c and some subset of A. Like provability, causation is monotonic and cumulative:

If $A \vdash c$ and B is a superset of A, then $B \vdash c$.
If $A \vdash c$ and $B, c \vdash d$, then $A, B \vdash d$.

The important part of the causal relation is that it captures the functional dependence of the domain variables. If we want to start the car, then we can turn the key when the battery is charged. On the other hand, having the car start when the key isn't turned or the battery is dead in some way violates the design specifications of the car, and will not cause the battery to be charged or the key to be turned. Of course, if we observe that the car is started, then we can infer that the key is turned and the battery charged, based on the knowledge of causation. But it is not possible to *plan* to change the state of the key by starting the car. This illustrates the difference between a causal relation and a merely correlational one. The causal relation is asymmetric: given that c causes d, it is not necessarily the case that d causes c; while if c is positively correlated with d, d is positively correlated with c.

These remarks leave open the question of whether, in a particular instance, it is possible to have a causation relation that is symmetric for two propositions, or more generally to have one that is *cyclic*, that is, contains a loop that leads from a proposition back to the proposition. Other commitments may answer this question: for instance, assuming that causes always precede their effects in time forces the causal relation to be acyclic. Without making any such commitments, we will assume that the causation relation is acyclic, so that if $A ==> c$ is true, then c cannot be a member of A.

There are some complications in defining a causal relation that we will mention here, without offering any definitive solutions. The first is that of inferred causation. In starting the car, one also starts a machine, because the car is a type of machine. If the proposition M stands for starting a machine, then it seems that we should also have $K, B ==> M$. The justification for this rule is based on the presence of $K, B ==> S$, and the relation between the propositions S and M. The essence of this relation is definitional: saying that S occurs is tantamount to saying that M occurs. Note that there can be other relations between two propositions in which one implies the other, but the implication is not definitional. A good example are correlational relations, which are discussed below. To accommodate inferred causation, we could either include an exhaustive list of the relevant one-step causes (such as $K, B ==> M$), or we could use a definitional theory to describe the relation between propositions, and generate the inferred causes from an initial set using the theory.

Another problem arises when our knowledge of the causation relation is partial. We have already remarked that we may not know all of the one-step causes, and thus the closed relation $==>$ may represent only a subset of the actual causation relation. Other kinds of uncertainty also exist.

For example, suppose we know that dialing the number "911" connects one
with either the police or the fire department, but we don't know which. The
action of dialing 911 is completely determinate, it's just that we don't know
the exact outcome. To express epistemic uncertainty of this kind, it is
necessary to describe the causation relation in an appropriate language. If we
let c stand for the action of dialing 911, d for calling the police, and e for
calling the fire department, then our knowledge is expressed by the statement:

$$Either\ c ==> d\ or\ c ==> e\ .$$

An interesting question is whether such statements can be reduced to simple
indeterminacy in causal statements, e.g., is the statement above equivalent to

$$c ==> (d\ or\ e)\ ?$$

In addition to causal relations, we often have information about co-
occurrences of events in a domain, without knowing whether one event
caused the other, or there is some underlying cause for both. This
information can be used to make predictions, but it does not contribute to
causal explanations. Correlations are represented by a first-order theory T.
Since all causation is also correlation (although not the converse!), for every
cause, we place its correlational equivalent in T. For example, if

$$B, K \longrightarrow S$$

is a one-step causal relation, then

$$B\ \&\ K -> S$$

is in T. This latter statement is a simple (material) conditional of the
propositional logic, saying that if both B and K are true, then so is S. Note
that, unlike the case with the causal relation, the material conditional can be
used in for "backwards" inference, e.g., if we know that S is not true, then
we can infer that either B or K is not true.

Besides correlations based on one-step causation, we may also have
correlations based on underlying but unknown causes, or on partial
information about causation. For example, we may know that the car will

not start when the battery is discharged, and we would add

$$D \to N$$

to the correlational theory. In addition, any definitional relations among propositions should be in the correlational theory. Since B and D are opposites, in T we have the statement:

$$B \leftrightarrow \neg D.$$

We can define a consequence operator for T, so that $A \vdash c$ just in case c follows from the set of propositions A and the statements of T. Because T contains all of the material implications corresponding to the causal relation, it is easy to show that the operator $\vdash$ is a superset of $\vdash:-$, that is

If $A \vdash:- c$, then $A \vdash c$.

III. NORMAL CONDITIONS

Suppose that we know the causation relation for some domain, and try to make predications from initial conditions. If our knowledge is complete, then we simply look at the transitive relation $==>$ and see what follows. But it is unusual to have such perfect knowledge. In starting the car, one does not *know* that the battery is charged, one simply assumes it because it is normally the case, and there is no reason to believe otherwise. One does not argue for the battery being charged on the basis of causal reasons: there may be such a reason, but it is unlikely that we know it. And the full battery is not a basic cause, since (at least for this example) we have no direct control over whether it is charged or not; if the car does not start because of a bad battery, we are frustrated just because it takes so much time to cause it to get back into working condition.

On the other hand, normal conditions are not part of a background "causal field" that is present but not reasoned about explicitly. Knowing reasons, causal or otherwise, for the *absence* of a normal condition can have an effect on causal reasoning. In the car example, knowing that the lights were on is a reason for disbelieving that the battery is charged, and influences our inferences about the effect of turning the key.

Formally, normal conditions are a distinguished set of propositions
N. They can be basic causes, but they need not be. For the car example, the
only normal condition is $N = \{B\}$, the battery being charged, which is not a
basic cause. In this example we have assumed that the only way the car does
not start when the key is turned is if the battery is bad, but in general our
knowledge of how things can go wrong is minimal. One way of
representing such incomplete knowledge is by employing a general
"normality" predicate. Suppose in the car example there were many ways in
which the car could fail to start; we could represent the normal causal relation
by

$$K, \; nS \longrightarrow S \; ,$$

where nS is a proposition asserting that the car is in working order. With
this generic "normality" predicate, we can specify that a bad battery causes an
abnormal condition by adding a one-step cause

$$D \longrightarrow aS \; ,$$

where aS, an "abnormality" predicate, is the complement of nS.
Abnormality predicates first appeared in McCarthy's formalization of
commonsense reasoning using defaults [1986]. Their use was almost
identical to that here: a device for stating a generic normalcy condition that
could be contraindicated by particular abnormal conditions.

In the car example, inserting the normal condition for starting yields
the diagram shown in Figure 2.

The normal conditions are indicated by ovals, and the primitive
causes by rectangles. Note we have added a normal condition for the lights to
cause the battery to die; this condition could be violated, for example, if the
lights were broken so that they did not use any electricity. The arrow
connecting *dead battery* with *normal start* has a circle through it, indicating
that it causes the complement of the normal condition.

As is usual with representational questions, there are alternative
ways to representing generic normal conditions and their particular causes.
We have chosen to consider a proposition like *dead battery* as a cause of the
generic abnormality, but we could just as readily have used a definitional
scheme in which a dead battery is considered as a type of abnormal start
condition. The effect would be the same, to introduce $L ==> aS$ as a member

of the causation relation.

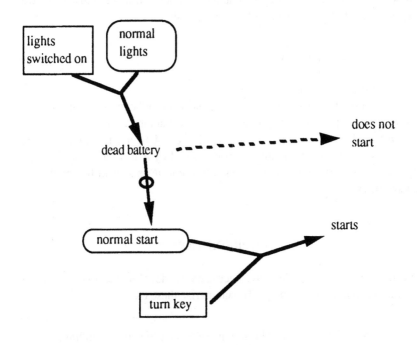

Figure 2. Starting the car

Normal conditions can also be introduced into the correlation relation, for example, we could state that normally A is positively correlated with B by adding the statement

$$nC \rightarrow (A \leftrightarrow B)$$

into the correlation theory T. Here nC is a normal condition associated with the correlation of A and B. One way to think of such correlations is as "soft" constraints among the propositions: normally one would expect them to be correlated, but there could be circumstances under which they would not be.

Soft correlations are a way of representing preferred choices for causation. For example, suppose there are three ways in which the proposition D could be cause: A, B, and C. Under normal circumstances, we expect that A will be the cause of D; so we write

$$nD \rightarrow (D \rightarrow A)$$

as part of the correlational theory.

In the next sections we consider how normal conditions are used in the inference operations of prediction and explanation. The concept of an *explained abnormality* will play a key role. Let A be a set of propositions; we will say that an abnormality aP is explained by A if A causes aP, that is, $A \mid:- aP$. The corresponding normal condition nP is said to be "explained away" by A.

IV. PREDICTION

We now have all the elements necessary to develop the two inference operations of causal reasoning. To summarize these elements:

1. A causal relation $==>$ and the corresponding consequence operator $\mid:-$.
2. A correlational theory T and the corresponding consequence operator $\mid-$. This consequence operator contains causal consequence.
3. A set of primitive causes C.
4. A set of normal conditions N.

These elements constitute a *default causal net*. In the car example, they are instantiated as follows.

One-step causes:
$$L, nL \longrightarrow D$$
$$K, nS \longrightarrow S$$
$$D \longrightarrow aS$$

Correlational theory:

$$L , nL \rightarrow D$$
$$D \rightarrow N$$
$$K, nS \rightarrow S$$
$$D \rightarrow aS$$
$$aS \leftrightarrow \neg nS$$
$$B \leftrightarrow \neg D$$
$$N \leftrightarrow \neg S$$

Primitive causes: $C = \{L, K\}$

Normal conditions: $N = \{nS, nL\}$

Within a default causal net, a *prediction* from a set of propositions A is simply any correlational consequence of A,:

c is predicted from A if $A \vdash c$.

Predictions tell us what to expect, given a set of initial conditions. Because causal consequence is a subset of correlational consequence, prediction includes both causal and noncausal reasoning. If we know the lights were on and they are working normally, then we can predict that the battery will be discharged; in this case, the initial conditions caused the predicted effect. On the other hand, given that we know the battery is charged, we can predict that either the lights were not on or they are not working normally. These conclusions are not caused by the initial conditions, although they are based on the causal relation.

In default causal nets, predictions are a weak form of inference, because they demand too much initial knowledge. In the car example, it is impossible to predict the car will start if the key is turned: one must also know that no abnormal conditions are present. This is where the normal conditions are useful: they are assumptions about how the world normally is that can be used to fill in gaps in our knowledge. Of course, such "filling in" must be consistent with what is already known. So we define an *extension* of a set of initial conditions A as follows.

> An extension of a set A is the union of A with a maximal set of normal conditions from N that are consistent with A (according to the correlational theory).

Here are two sets of initial conditions and their corresponding extensions, along with predictions from the extensions.

Initial conditions	Extensions	Predictions
K	K, nS, nL	S, ¬N, ¬L
L	L, nL	D, ¬B, aS, ¬nS
	L, nS	¬nL, ¬D, B
L, K	L, K, nL	D, ¬B, aS, ¬nS, N, ¬S
	L, K, nS	¬nL, ¬D, B, S, ¬N

In the first instance, we know that the key is turned but nothing else. It is consistent to assume all normal conditions, and so there is only one extension, in which the car starts. Note that, because of the correlational theory, it is also possible to predict that the lights were not switched on, since if they had been, the car would not have started. So the assumption of normal conditions may mean that some primitive causes are predicted not to have occurred.

In the second instance, we know that the lights were switched on but nothing else. Now there are two maximal sets of normal conditions that are incompatible, since L, nL imply the negation of nS. If we assume the lights were working normally, then we predict the battery is dead; otherwise, assuming the car is in working condition for starting, the battery will be ok and something is wrong with the lights. Both these cases are logical possibilities, yet intuitively we prefer the first extension. It seems more reasonable to assume that, if the lights were switched on, the battery would be discharged and so the car would not work. The abnormality nS can be explained by the lights being on (assuming the lights are working correctly), but not the other way around: the car being in working order does not cause the lights to malfunction, and hence does not explain nL.

A similar analysis holds for the third instance, when both the lights are switched on and the key is turned. Again there are two extensions, since the normal conditions are not compatible. In one extension the car does not start because the battery is dead, and in the other the car starts, and the lights are not working. As in the second instance, the more reasonable prediction is that the car will not start.

How can we eliminate extensions that do not lead to intuitively correct predictions in the causal theory? Note that the unwanted extensions are bad only in comparison to other extensions of the initial conditions. For example, if we add to the third instance the information that the battery is charged, then the only possible extension is the one in which nS holds. So unwanted extensions are found by comparison to other, more reasonable ones. The distinction lies in which normal conditions are chosen: we prefer extensions that have fewer unexplained abnormalities. Here the concept of "explanation" is the technical one of the last section: the absence of the normal condition nS is explained by the extension L,nL because its negation is caused by these propositions.

To formalize the preference for "good" extensions, let us define the *adjunct* of a set of propositions A as all normal conditions that are not either in A or explained by A. The adjunct characterizes the unexplained abnormalities of A. Here are the adjuncts of the extensions in the last two instances of the previous example.

Initial conditions	Extensions	Causal consequences	Adjuncts
L	L, nL	D, aS	*none*
	L, nS	*none*	nL
L, K	L, K, nL	D, aS	*none*
	L, K, nS	S	nL

The first extension of each instance has an empty adjunct set, because the only normal condition not included in the extensions, nS, is explained by the extension. But nL is not explained by the second extension in each instance, and is part of the adjunct. We prefer extensions that have minimal adjuncts, relative to other extensions.

> An extension E of a set of propositions A is *ideal* if there is no other extension E' of A whose adjunct is a proper subset of the adjunct of E.

A proposition P is *ideally predicted* from a set A if it is predicted in every ideal extension of A. We take the ideal predictions of A to be the consequences of *A* that it is reasonable to believe in. For the car example above, there is only one ideal extension for each set of initial conditions, and, as expected, we would ideally predict that the car does not start when the lights were on and the key is turned.

V. EXPLANATIONS AND EXCUSES

Predictions tell us what to believe, given a set of initial conditions. Often we are also interested in how a set of observed conditions could have arisen, that is, we seek an explanation of the observed state of affairs. Explanations are a powerful means of deriving further information about the world. For example, if we learn that the car started, then the only causal explanation of that happening is that the key was turned and the car is working. From this we can predict other consequences, e.g., the battery is charged, and so on

The question of what constitutes an explanation for a set of observed events is a complicated one, and like formal accounts of causation, there are different approaches to explanation, some based on conditional probabilities [xxx], others on more refined theories of causation [Dretske]. One of the important properties of explanations, no matter what the formal account, is that they are task-sensitive. For example, take the question of whether normal conditions are considered to be part of the explanation. In the car example, suppose we observed that the car started; it may be enough of an explanation to say that the key was turned, without mentioning the normal condition that the car is working, or that other conditions such as the battery being charged indirectly contribute to the car's normal functioning. But if the car does not start when the key is turned, we may be interested in how the normal conditions were violated: it is important to know that the battery is discharged, so we can fix it. And if we are interested in preventing a recurrence of the observed behavior, we may also look for the reason that the battery is discharged: that the lights were on.

Most explanatory tasks require that the explanation be causally related to the observations; this is the point where probabilistic approaches tend to suffer. Prediction from a set of conditions is not the same as explanation using these conditions. It is plausible to predict that the lights were not on if the car starts when the key is turned. But if we try to explain why the lights were not on, it does not sound reasonable to say, "because the car started when the key was turned." Explanations involving a causal connection to the observations seem to be required, mostly because the underlying task involves fixing or changing the observed conditions.

Given these remarks, and the limitations they entail, we will give a simple formal theory of explanations within default causal nets. The causal connection between explanations and observations is obviously available by means of the causation relation. More difficult is deciding which propositions that causally imply the observations actually constitute an explanation. Assuming that the most likely use for explanations is to diagnose and fix a system, we will let explanations be sets of primitive causes, since they are the propositions over which we have direct control.

> Let O be a set of observations. An *explanation of O* is any set A of primitive causes such that $A \mathrel{|:-} O$.

This definition is similar to that for prediction, but employs the causation operator $|:-$ instead of the correlation operator $|-$, in keeping with the concept of explanations as causal relations. As might be expected from the comments on prediction, this definition of explanation is too weak to be of much use unless normal conditions are introduced as defaults. In the car example, there is no explanation for why the car starts, since the normal condition nS is not a primitive cause, and there is no causal relation that leads to it.

Similarly to what we did for predictions, we can remedy the weakness of explanations by considering the addition of normal conditions to form extensions.

> An *O-extension of a set A* is the union of A *with* a maximal set of normal conditions from N that are consistent with A and O (according to the correlational theory).

The only difference is that consistency is with respect to both A and O, rather than just A. Now we define a normal explanation:

> A *normal explanation of O* is a set of primitive causes A with an O-extension E such that $E \mathrel{|:-} O$.

Normal explanations are a much more powerful means of inferring causation; in the car example, we can explain the car starting by the normal explanation

$A = \{K\}$, since it has the extension $\{K, nS, nL\}$ in which the car is caused to start.

To take a slightly more complicated example from the car domain, assume that if the light bulbs are removed, turning the lights on does not cause the battery to discharge. A one-step causal rule takes care of this:

$$R \longrightarrow aL ,$$

where R stands for the bulbs being removed, and aL is the abnormality complement to nL. Now assume that we observe that the light switch was turned on, and yet the car starts, so that the observation set is $O = \{L, S\}$. (Note that there is no restriction against observations being primitive causes; it just means that they will also appear in any explanation.) The following table contains the normal explanations of O, along with their extensions and causal consequences.

Normal explanation	O-extensions	Causal consequences	Predictions
K, L	K, L, nS	S, L	$B, \neg D, \neg N, aL, \neg nL$
K, L, R	K, L, R, nS	S, L, aL	$B, \neg D, \neg N, aL, \neg nL$

The predictions of an O-extension are defined as usual from the correlational theory. There are two normal explanations: in the first, the key is turned and the light switch is on, but for an unexplained reason the lights are not working and the car starts normally. This is reflected in the O-extension, in which nS appears, but not nL. For the second explanation, the key is turned and the lights switch is on, but the light bulbs are removed. In this case, the abnormality condition aL is a causal consequence of the O-extension, so that the normal condition nL is denied, and the battery is not discharged. By default, turning the key starts the car.

Both these explanations are formally correct, given the assumption of normal conditions, but there is a sense in which the second one is preferred, since it has fewer unexplained abnormal conditions. It is a "more normal" explanation. As with predictions, we can translate this preference into a formal criterion by defining an ideal explanation using the adjuncts of the O-extensions. Recall that the adjunct of a set of propositions are all those normal conditions that are excluded from the set, and which are not explained by the set.

> A normal explanation A of O is ideal if there is no other normal explanation for O with an O-extension whose adjunct is a proper subset of the adjunct of the O-extension of A.

The definition of ideal explanations may seem a little complicated, but it really depends on two very simple operations. First, find the normal explanations of a set of observations, by looking at the O-extensions of sets of primitive causes. Then, among these extensions, delete the ones that have too many unexplained normal conditions, relative to the others. The ones that are left belong to the ideal explanations. In the car example, $\{K, L, R\}$ is ideal because the adjunct of its O-extension is empty, whereas $\{K, L\}$ has the unexplained abnormality aL. As nature abhors a vacuum, so to we, as ideal reasoners, abhor explanations with abnormal conditions that cannot be explained.

One remark about a point mentioned above, including primitive causes in the observations: it makes more sense to think of observed primitive causes as being part of the background theory, since they do not need any explanation. In fact, other observed propositions that are not primitive causes might also be considered as not needing explanation and included with the theory, depending on the explanation task. For example, we may know that the battery is dead, and not be interested in why this is so, but only whether the car will start. So in a general explanation task, we could divide observed propositions into a set of "givens" that need no explanation but are included in the theory, and a set of propositions to be explained.

Explanations draw a strong connection between primitive causes and observations; for some tasks, it is useful to draw a weaker connection, which we call an *excuse*. Suppose in the car example we observe that the car does not start even though the key is turned. There is no causal connection in the theory to the proposition N, and so there can be no explanation of the observations. But intuitively there seems to be something that needs explaining: why didn't the car start, since that would normally happen when the key is turned? Without further assumptions, given the key turning, we would ideally predict the opposite of the observation N. An excuse is a set of primitive causes that, if assumed, would erase this contradiction between prediction and observation.

We can define excuses formally in terms of prediction.

> An ideal excuse for an observation set O is a minimal set of primitive causes that does not ideally predict a contradiction of O.

For the example just mentioned, we are given that the key is turned, and looking to excuse the observation N. The only possible excuse is L, the lights were switched on, since this will cause the abnormal condition aS and block the prediction of S. Excuses are useful in diagnosing and repairing systems that deviate from their normal behavior. If we want the car to start when the key is turned, then we shouldn't leave the lights switched on.

Explanations and excuses do not exhaust the ways we can use default causal nets to understand and predict the behavior of causal systems. To mention one other possible task: suppose we are interested in preventing a particular proposition from coming about; for concreteness, we are trying to prevent the battery from being discharged. If we remove the bulbs from the lights, then no matter what further primitive causes occur, the battery will not be discharged. So one way of viewing the task of prevention is to find a set of primitive causes that can be performed to make it impossible for the undesirable proposition to be caused. This notion of prevention is a simple one and needs refinement (at least in terms of a distinction between primitive causes that are under our control, and those events that may occur based on the actions of other agents), but it points out how useful the concepts of causation and normal conditions are in a different setting.

VI. AI THEORIES OF DEFAULTS AND CAUSATION

The theory of default causal nets provides a formal mechanism for commonsense reasoning about causation and correlations in everyday domains. We have argued that the reasoning processes of prediction and explanation make essential use of normal conditions to bridge gaps in knowledge, and enable us to draw conclusions from imperfect information. The importance of the formalization is the precision it gives us in analyzing the concepts. The tools of logical language and consequence operations are the most natural way to accomplish this task, and the availability of proof methods means that the inferences can be mechanized, an important step to providing computer agents with the same commonsense capabilities that we possess.

In developing the theory, we have relied heavily on knowledge representation techniques developed in AI, especially the concept of default or normal condition. It was recognized early, in the work of McCarthy [1980], that commonsense reasoning was fundamentally different from deductive logic in that commonsense inferences could not be deductively substantiated given the partiality of information available to a reasoner. We tend to "fill in the gaps" of our knowledge in reasonable ways, making assumptions about the world that are often falsified; yet for the most part they enable us to successfully predict the future course of events and the effects of our own actions. The response of AI was to develop formal techniques for making default assumptions, and the resultant formal systems have become known as *nonmonotonic logics* (in contrast to deductive logic, which is monotonic: adding more information does not change previously drawn conclusions). The technique of specifying a set of normal conditions, and defining extensions as a maximally consistent augmentation using these conditions, was originated by Reiter [1980] and developed by many others, especially Poole [1990], whose THEORIST system contains many of the prediction and explanation inference techniques presented here. By now the study of nonmonotonic logic is a cottage industry in AI, and contributes to some of the most sophisticated mathematical reasoning in the field.

While the mathematical study of default reasoning has blossomed, its application in domains such as reasoning about events and their causal relations has received less attention; it was as if everyone assumed that once the technical problems were solved, the applications would be trivial. Would that it were so! In a study of the application of nonmonotonic logic to the problem of reasoning about actions, Hanks and McDermott [1987] showed that every major formal technique had serious problems in representing even simple commonsense reasoning for this domain. The problems were not with the logics themselves, but rather, the domain was more complex than had been realized, and the necessary concepts and relations had not been developed; the nonmonotonic logics, like first-order logic, are meant to characterize general properties of reasoning, and do not provide any information about domain-specific properties and relations. And defining the correct concepts and their interrelations has undeniably proven to be a difficult task.

The theory we presented here, default causal nets, is a contribution in the application of default reasoning techniques to a commonsense reasoning domain. It develops the concept of an irreducible causation relation, and joins to this the ability to make assumptions about normal conditions. The interaction between causation and assumption is complex; the most interesting facet is in the inference processes of prediction and explanation, where there is a bias towards assuming as few unexplained abnormalities (those with no causal account) as possible.

The inspiration for using a primitive causal relation comes from the work of Pearl [1988], who introduces causation as a structuring concept in Bayesian inference. In fact, he introduces a graph structure called a Bayesian net, in which the causal connections between propositions are explicitly represented. The causal relation, as here, is primitive; it is not defined in terms of conditional or any other kind of probabilities. In the same way, in a logical treatment, we define the causation relation as primitive, rather than reducing it to material implication or some other logical relation. By analogy with Bayesian nets, we have called the causal relation a causal net. Causal nets differ from Bayesian nets in the obvious fact of being qualitative rather than quantitative, and relying on logical inference rather than computing posterior probabilities. The normal tradeoffs between probability theory and logic are present: Bayesian nets demand more information in the form of marginal and prior probabilities, but give a finer distinction in the truth-values of propositions, and gradations of strength in the causal connections between propositions. One of the advantages of introducing normal conditions is to import some of this representational flexibility into a logical structure: for instance, it is possible with normal conditions to state that one way of causing a given proposition is preferred over others.

Our emphasis on prediction and explanation is also the main concern of an area in AI called model-based diagnosis, and we have incorporated several ideas from here. In its standard formulation [Reiter 1987], a system to be analyzed is considered to be composed of a finite number of components, each of which can be functioning normally, or be broken. A first-order theory states the relation between the input and output of each component, assuming that it is working; it also describes the connection between components. Given a set of inputs, and assuming that all components are working, the theory predicts what the outputs will be. If this prediction differs from the observed outputs, a minimal set of normality assumptions are retracted, until the predicted effects no longer conflict with observations; the resulting abnormal components are a diagnosis of the observed output. Note the similarities between this and our definition of an excuse. Normally functioning components correspond to normal conditions, and the system theory is our correlational theory. A diagnosis of the observed output is just an excuse, that is, a maximal set of normal conditions consistent with the output. What the model-based theories lack is the concept of causation, and the corresponding notion of explanation. Recently there have been proposals introducing so-called "fault models" into the theory, describing the abnormal functioning of components [e.g., de Kleer 1990]. These changes are a first step towards incorporating explanations; but there is still no explicit notion of causation. We believe that the only way to fully represent the complications of the interaction of defaults and causation is, as here, to admit causation as a full-fledged concept into the theory.

There is another area of AI that has influenced this research, which is concerned with reasoning about action. It was recognized by McCarthy and Hayes [1979] that if actions were represented in a natural way, by stating the connections between situations obtaining before and after an action was performed, then there is a representational problem, which they named the "frame problem." Simply put, if we describe an action compactly by stating the changes it brings about, how do we state the implied consequence that everything not affected by the action remains the same? The straightforward nonmonotonic logic approaches to the problem, as Hanks and McDermott showed, gave counterintuitive results. One solution was to introduce causation, or concepts that were similar in structure [Lifschitz 1989, Shoham 1988]. The concern of these theories has been different from ours, since they concentrate on solutions to the frame problem, while we are more interested in the interaction of causation, correlation, and normal conditions. Still there are many points of correspondence: one could consider solving the frame problem by using normal conditions, for instance.

Finally, there are some recent theories that, like ours, take their inspiration from Bayesian nets and attempt to incorporate both causation and normal conditions in a logical theory. The most advanced example of this is in Geffner's theory of causal and conditional reasoning [Geffner 1989]. Our concept of an adjunct to an explanation, and the preference for explanations that have a minimum of unexplained abnormalities, are also key aspects of Geffner's theory, and his work has influenced our understanding of the interaction of causation and normal conditions.

BIBLIOGRAPHY

de Kleer, J. and Williams, B. C.: 1990, Diagnosis with behavioral modes. *Proceedings of the National Conference on AI*, Boston, Massachusetts.

Geffner, H.: 1989, *Default Reasoning: Causal and Conditional Theories*. PhD thesis, Department of Computer Science, University of California at Los Angeles.

Hanks, S. and McDermott, D.:1987, Nonmonotonic logic and temporal projection. *Artificial Intelligence* **33** (3).

Lewis, D.: 1973, Causation. *Journal of Philosophy* **70**, pp. 556-567.

Lifschitz, V.: 1989, Miracles in formal theories of action. *Artificial Intelligence* **38** (2).

McCarthy, J.: 1980, Circumscription --- a form of nonmonotonic reasoning. *Artificial Intelligence* **13** (1-2).

McCarthy, J.: 1986, Applications of circumscription to formalizing commonsense knowledge. *Artificial Intelligence* **28**.

McCarthy, J. and Hayes, P.: 1979, Some philosophical problems for the standpoint of Artificial Intelligence. In B. Meltzer and D. Michie, eds., *Machine Intelligence* **9**, Edinburgh University Press, pp. 120-147.

Pearl, J.: 1988, *Probabilistic Reasoning in Intelligent Systems: Networks of Plausible Inference*. Morgan Kaufmann.

Poole, D.: 1990, A methodology for using a default and abductive reasoning system. *International Journal of Intelligent Systems* **5** (5), pp. 521-548.

Reiter, R.: 1980, A logic for default reasoning. *Artificial Intelligence* **13** (1-2).

Reiter, R.: 1987, A theory of diagnosis from first principles. *Artificial Intelligence* **32**.

Suppes, P.: 1970, *A Probabilistic Theory of Causation*. North Holland.

Shoham, Y.: 1988, *Reasoning about Change*. MIT Press, Cambridge, Massachusetts.

Cohen, P.R., Morgan, J., and Pollack, M. E., (eds).: 1990, *Intentions in Communication*. MIT Press, Cambridge, Massachusetts.

SRI International
333 Ravenswood Ave
Menlo Park, CA 94025
konolige@ai.sri.com

JEFF PARIS AND ALENA VENCOVSKA[*]

PRINCIPLES OF UNCERTAIN REASONING

The aim of this paper is to discuss some principles or assumptions arising in uncertain reasoning, and to investigate their relationships and consequences within the particular framework of probability logic. In order to motivate this framework and to provide a context in which to judge these principles we first consider a simple example.

I. EXAMPLE E

Suppose that we are interested in constructing an expert system, say for diagnosing patients who turn up at a health centre. That is, we want to develop a computational device which, when given as input the presence, absence, severity of the various signs, symptoms, features etc. applying to the patient outputs an acceptably accurate qualified diagnosis. Arguing that a doctor himself appears to be just such a device might lead us to be optimistic that this might be possible, provided that we can build enough knowledge and expertise into our expert system.

From discussions with a doctor we are lead to understand that in the course of his work he has acquired a number of pieces of general knowledge which he considers relevant to this problem of diagnosis and these he freely relates to us. In this way we obtain a set, K_0 say, of general knowledge statements such as:

Symptom D strongly suggests disease B

Symptom D is rather uncommon

A patient of type E cannot have disease B

About 50 % of patients are of type E

Condition F is mainly found in patients of type E

etc. etc.

A. Clark et al. (eds.), Philosophy and Cognitive Science, 221–259.

Having spent a sufficient length of time collecting K_0 we might now feel that we can construct a reasonable expert system by simply reading the answers out of K_0. For example if we inputed just

Patient complains of symptom D

the device would respond

Disease B strongly suggested

by directly using the corresponding ready made 'rule' in K_0.

The trouble with this expert system is that it will only work if our input corresponds to the antecedent of a rule in K_0, as in the above example. But what if instead we knew (just) that

Patient complains of symptom D and exhibits condition F

and we do not have a rule in K_0 linking this antecedent directly to the diagnosis? The absence of any such rule might lead us to conclude that, despite the time already taken, K_0 was still not a good representative of the doctor's knowledge, after all the doctor would not get stuck here, he would be able to pull out a suitable rule. So perhaps we should go back and directly ask the doctor for rules to cover all such eventualities.

However it should be clear that it is completely infeasible to do this since in the course of collecting K_0, even for quite simple expert systems, the doctor has probably mentioned at least 40 signs, symptoms etc. so that asking the doctor for rules to cover all such combinations would require him to produce at least 2^{40}, around a million million, such rules.

Clearly then we had better come up with a method of producing such rules ourselves on the basis of what we do have i.e. K_0.

Furthermore we might well be optimistic that this is possible. After all the doctor surely cannot have all these rules already written down in his head, he must be generating them in some way from what he does know and hence, since he has endeavoured to embody in K_0 what he believes to be relevant, K_0 should surely suffice as a basis for us to generate, at least a reasonable approximation to, the rules the doctor would have given. The question is, how should we do it?

Similarly we might ask the same question not about unspecified rules but about unspecified base rates. For example the doctor can, apparently without much difficulty, say how likely he thinks it is for a patient to suffer from B and F. Arguing as above we might again feel that an approximation to this likelihood (we use such words here in a non-technical sense) should be derivable from K_0. The question again is how?

Of course the reader might object to the above argument. Whether or not one accepts it the problem of constructing an expert system from K_0 remains the same. However in the absence of such an argument there seems no reason, per se, why our endeavour should, or could, succeed.

In this example E, the inductive inference, or as we shall call it *uncertain reasoning*, which we seek amounts to nothing more than a method for predicting the doctor's answer to a question given some information K_0 about him. Interesting, and practically important, as this may be however it is not our wish to limit ourselves to such a singular example of inexact reasoning (although of course many of our ideas and stimuli will arise from considering ourselves). We shall rather consider the more general problem of predicting the answers of an 'intelligent' agent, or expert (but not necessarily human) given only some fragment K of the agent's general knowledge. Thus we shall be interested in the *possibilities* for uncertain reasoning, arising out of making various mathematical assumptions, which might be considered reasonable for an 'intelligent' agent, although perhaps not holding for real human agents. (Hence the reader might find that this text makes more sense if expressions like 'doctor', 'agent' are interpreted, as we intend, in this wider sense). We shall pursue this interest by presenting some, popular, modes of uncertain reasoning which have been proposed for answering the sort of problem raised in example E and will concentrate in particular on the mathematical assumptions underlying them, and the justifications and consequences of these assumptions.

It is important here to emphasise that we are interested in the problem of predicting the expert's answers to questions. For in many such situations there are also notions of 'correct' answers. For example in E the actual proportion of patients suffering B and F at similar clinics in the past year could be taken as a 'correct' answer to the question of how likely it is that a patient will be suffering from B and F. Presumably the prediction of some such correct answers is the primary aim of expert system builders. However as we said our target will be the expert's answer whether or not it agrees with 'the correct answer'.

One could, of course, question here why there should be any limit on the possibilities for uncertain reasoning, why any one mechanism, no matter how silly, would not be just as acceptable for generating answers given, say, $\mathbb{K}_0$ as in example $\mathbb{E}$, as any other. In reply to this, however, we observe that this would, for example, allow us to conclude from $\mathbb{K}_0$ that

Symptom D is a certain indicator of patient being of type E

However this conclusion appears inadmissible since it directly contradicts $\mathbb{K}_0$. Similarly the conclusion, based on the displayed first five statements in $\mathbb{K}_0$, that

Patients complaining of symptom D invariably have condition F

whilst not directly contradicting these first five statements would certainly seem to fly in the face of reasonableness or common sense. Thus requirements of consistency and common sense, more grandly termed 'principles of uncertain reasoning', certainly seem to provide some sort of limits on the possibilities for uncertain reasoning.

Additional limits could also be argued for on other grounds, but we only briefly touch on them here. First we need to formulate the general problem suggested by example $\mathbb{E}$ in a mathematical setting.

II. MATHEMATICAL FORMULATION

We shall work within the following, limited, formulation of the problem. We shall assume that the general knowledge statements given by the expert (e.g. doctor in example $\mathbb{E}$ can be written in the form of a set of equations involving a belief function defined on sentences of a propositional language. This is best explained by an example. Consider the statement from example $\mathbb{E}$,

Symptom D is rather uncommon

We might express this as

$$Bel_0(D) = 0.01$$

where $Bel_0(X)$ stands for 'the doctor's belief that a random patient, or more concretely *the next patient through the door*, will have property X' and 0.01 is a number on the scale [0,1] corresponding to 'rather uncommon' in this context, where $0 \equiv$ 'never', $1 \equiv$ 'always (or certain)' and $\frac{1}{2} \equiv$ 'no preference (or indifference)'.

Similarly the knowledge statement

Symptom D strongly suggests disease B

might be expressed as

$$Bel_0 (B|D) = 0.9$$

where $Bel_0 (X|Y)$ stands for the doctor's 'belief that the next patient through the door will have property X, given that he has property Y' and 0.9 corresponds to 'strongly suggested' on the same scale [0,1].

More formally then, we are treating the signs, symptoms diseases etc. such as B,D,E,F as propositional variables in some finite propositional language L and

$$Bel_0 ( ) : SL \rightarrow [0,1]$$

$$Bel_0 (|) : SL \times SL \rightarrow [0,1]$$

where SL are the sentences of L (formed by repeated application of the connectives $\wedge$ (and), $\vee$ (or), $\neg$ (not) to the propositional variables). We shall refer to $Bel_0 ( )$ as a *belief function* and $Bel_0 (|)$ as a *conditional belief function*, and their values in [0,1] as belief values. (Various other names are used in the literature, for example truth values, certainty factors, confidence coefficients). We shall assume for the present that in this way then $\mathbb{K}_0$ of example $\mathbb{E}$ can be re-expressed as a finite set of identities over the reals, $\mathbb{R}$, involving Bel_0 (θ) and further identities involving Bel_0 ($\phi \mid \psi$) for some θ, ϕ, $\psi \in$ SL.

A few comments are in order here. Firstly why are we justified in using real numbers in [0,1] to represent belief values? Well, in a sense, this does not *yet* need any justification because we could simply claim that 0.01, say, is just a code for 'rather uncommon' so that $\text{Bel}_0(D) = 0.01$ is just a coded version of 'Symptom D is rather uncommon' and nothing has been lost. However it is clear that this defence will not survive long since we obviously shortly intend to apply arithmetic relations and operations to these numbers and hence, in effect, between their decoded versions. In particular the key use of this will be in the linear ordering so that, for example, it will be implicit that 0.01 ($\equiv$ rather uncommon) expresses less belief than 0.9 ($\equiv$ strongly suggested) which itself is less than 1 ($\equiv$ certain).

Indeed it may be that this assumption, that beliefs can be ordered like the reals in [0,1] may be sending us on the wrong path (what would that mean?) or at least unduly limiting the possibilities we spoke of earlier. However, in favour of this assumption it does seem that in practice the expert is able, when asked, to express his beliefs on this scale, whereas we see no more attractive setting for belief values. Indeed in the formalisation of our example $\mathbb{E}$ (and others like it) we shall assume that the figures given are the ones the doctor, or expert, would himself match with his verbal expressions of qualified belief.

A second objection here, perhaps, to this formalisation is that this is far from clear that the doctor's (or in general, expert's) beliefs are all of the same nature or quality, for certainly they might well arise from various origins. For example, whilst some of the doctor's beliefs might seem to result from some subconscious, personal, statistical data, it is surely the case that others have, at least partly, been influenced by reading books, by analogy, by symmetry, personal prejudices etc. etc.. (Indeed although use of the terms 'doctor' and 'expert' might suggest that they possess some sort of divine authority we wish to dispell here and now any misconception on this point. The term 'expert', or a synonym for it, will be used simply for the agent who provides the knowledge statements. We do not assume any objective truth for these statements.) Of course the fact that beliefs may come from various sources does not of itself imply that there are various different sorts of belief, just as iron is the same thing no matter where it is mined or how it is extracted, although it is certainly a point which we may well be forced to reconsider in the future. For this paper however we shall try to simplify matters as much as possible treating all beliefs as being of the same quality and nature.

III. THE CENTRAL QUESTION

Referring still to example $\mathbb{E}$, $\mathbb{K}_0$ has now been formalised as a finite set of identities of the forms

$$\text{Bel}_0(\theta) = a, \qquad \text{Bel}_0(\phi|\psi) = b,$$

for some θ, ϕ, $\psi \in SL$. Here $\text{Bel}_0(\ )$ is the doctor's belief function, $\text{Bel}_0(\ |\ )$ his conditional belief function and these identities represent all we know about these functions. What we would like is to be able to predict the values of $\text{Bel}_0(\chi)$, $\text{Bel}_0(\chi \mid \eta)$ for some other χ, $\eta \in SL$.

Rephrasing this slightly let K_0 be the result of replacing the doctor's actual belief functions $\text{Bel}_0(\ )$, $\text{Bel}_0(\ |\ )$ throughout $\mathbb{K}_0$ by function variables $\text{Bel}(\ )$, $\text{Bel}(\ |\ )$ standing for a belief function and a conditional belief function respectively. Then our problem can be expressed as: what values to give to $\text{Bel}(\chi)$, $\text{Bel}_0(\chi \mid \eta)$ on the basis of the set of constraints K_0, for χ, $\eta \in SL$? Notice that $\text{Bel}(\ ) = \text{Bel}_0(\ )$, $\text{Bel}(\ |\ ) = \text{Bel}_0(\ |\ )$ is a solution of K_0 and that this sums up all we know about $\text{Bel}_0(\ )$ and $\text{Bel}_0(\ |\ )$. This question is a special case of the main question around which this paper will be centred, viz,

Q   *Suppose K is a finite consistent set of linear*
        constraints over $\mathbb{R}$,

$$\sum_{j=1}^{r} a_{ji}\text{Bel}(\theta_j) = b_i \qquad i = 1, ..., m,$$

$$\sum_{j=1}^{r'} a'_{ji} \text{Bel}(\theta_j | \phi_i) = b'_i \qquad i' = 1, \ldots, m',$$

*for some θ_j, $\phi_i \in SL$. Then on the basis of K
what value should be given to Bel(θ), Bel($\theta | \phi$)
for θ, $\phi \in SL$?*

Here L is a finite language for the propositional calculus and Bel( ), Bel(|) are function variables standing, respectively, for a belief function and a conditional belief function, that is functions from SL into $[0,1]$ and from $SL \times SL$ into $[0,1]$. (Of course in the rest of this paper we shall often blur this distinction between a variable standing for a function and an actual function, for example using Bel in both roles).

Several points about this question Q require explanation. Firstly *consistent* here means that, with additional conditions on Bel( ), Bel(|), to be specified shortly, there is a solution satisfying these conditions and the equations in K.

Secondly we have generalised K beyond the form (assumed) of K_0 in example $\mathbb{E}$ by allowing linear constraints rather than the simple identities that, in practice, arise naturally. As we shall see shortly however, further conditions which we shall consider imposing on Bel( ), Bel(|), will naturally give rise to such generalised constraints, so that mathematically it is more elegant to start with this general form. (It is true of course that this might complicate the derivation of the 'answers' Bel(θ), Bel($\theta | \phi$) we might propose, and indeed many practical procedures do limit themselves to the former, simpler, K.) Of course our K could be generalised still further by allowing polynomial, rather than just linear, constraints, indeed the reader might have felt that our original natural language knowledge statements in example $\mathbb{E}$ could conceivably have yielded such constraints. However this introduces some significant new difficulties, so on the basis that we already have problems enough, we shall resist the temptation. Other possible generalisations here would be to allow inequalities and or constraints mixing conditional and unconditional beliefs. However these seem rather unnatural to us so we shall stick for the present with the above question Q.

Finally, and most importantly, in question Q the word 'should' is not intended to suggest necessity but rather implores a search for justifications.

Unfortunately as things stand it would be very exceptional if the set K of constraints in question Q determined Bel(θ) or Bel($\theta \mid \phi$) without any further assumptions. Thus we need in general to invoke some additional assumptions and considerations.

As a first step assumptions frequently involve properties of Bel( ) and Bel(|) The assumption which we shall make throughout this paper is that:

Bel( ) is a probability function. That is for $\theta, \phi \in$ SL,

(P1)   if $\models \theta$ (i.e. θ is a tautology) then Bel(θ) = 1,

(P2)   if $\models \neg (\theta \wedge \phi)$ (i.e. θ and ϕ are exclusive) then
               Bel($\theta \vee \phi$) = Bel(θ) + Bel(ϕ)

and Bel(|) is the corresponding conditional probability, that is,

$$Bel(\theta \mid \phi) = \frac{Bel(\theta \wedge \phi)}{Bel(\phi)} \text{ for } Bel(\phi) \neq 0.$$

Thus (P1) says that tautologies get belief value 1 whilst (P2) says that if θ and ϕ cannot both be satisfied at the same time then the belief value of $\theta \vee \phi$ is the sum of the belief value of θ and the belief value of ϕ.

There are several arguments based on common sense, or rationality, justifying the assumption that *Bel* is a probability function. One particularly appealing argument emerges from some early work of Ramsey, de Finetti, Kemeny, Shimony and Lehman (see [13], [5], [7], [14] [8]).The idea is to identify an expert's belief Bel(θ) in $\theta \in$ SL with his willingness to bet on θ being true in the real world, that is on $V(\theta) = 1$ where $V : L \rightarrow \{0,1\}$ is the valuation which represents the true (but possibly unknown) state of the world.

To be precise suppose that $0 \leq p \leq 1$ and that the expert is required to make a choice, for stake S > 0, between

(i)   gaining $S(1—p)$ if $V(\theta) = 1$ whilst losing Sp if $V(\theta) = 0$,

(ii)   losing $S(1—p)$ if $V(\theta) = 1$ whilst gaining Sp if $V(\theta) = 0$.

Clearly if $p = 0$ he could not do better than to choose (i) whilst if $p = 1$ he could not do better than to choose (ii).

Furthermore if $0 \leq p' < p \leq 1$ then it would be irrational of the expert (i.e. against his best interests) to choose (i) for p and (ii) for p'. For suppose he was to make such choices. Then he stands to gain strictly more by choosing (i) at p' than he did by choosing it at p (since he would gain $S(1—p') > S(1—p)$ if successful and lose $Sp' < Sp$ if not). However he picked (ii) at p' so, rationally, he must believe that he stands to gain at least as much by picking (ii) at p' than (i) at p' and hence strictly more than (i) at p. But then, as above, since $p > p'$ he stands to gain more by picking (ii) at p than he did by picking (ii) at p' and hence by picking (i) at p. But this contradicts the rationality of his choice of (i) at p.

From this it follows that if β is the least upper bound of the set of $p \in [0,1]$ for which the expert prefers (i) then for any $0 \leq p < \beta$ he prefers (assuming as we do that he is rational) (i) at p and for any $\beta < p \leq 1$ he prefers (ii) at p.

Thus β could be said to measure the expert's willingness to bet on θ, since if $0 \leq p < \beta$ he prefers a bet which pays him if θ is true whilst if $\beta < p \leq 1$ he prefers a bet which pays him if θ is false. Identifying belief with willingness to bet then leads to identifying $Bel(\theta)$ with β.

In a similar fashion we could argue for identifying the expert's conditional belief, $Bel(\theta \mid \phi)$, with his willingness to bet on θ on condition that ϕ holds. (If ϕ does not hold then the bet is null and void).

Now suppose that $Bel(\theta), Bel(\theta \mid \phi)$ are defined in this way for all θ, $\phi \in SL$. Then it could be said that these values are rational or *fair* if it is not possible for an opponent to arrange a Dutch Book against him, that is arrange a finite sets of bets (for various stakes) each of which the expert would agree to but whose combined effect would be to cause him certain loss no matter what V is.

Thus the assumption of rationality leads to the conclusion that the assigned truth values are fair. But as the next theorem (see above references) shows this is equivalent to *Bel* being a probability function.

Theorem 1 *If the values* $\text{Bel}(\theta)$, $\text{Bel}(\theta \mid \phi)$, θ, $\phi \in SL$ *are fair then Bel is a probability function and*

$$\text{Bel}(\theta \mid \phi)\text{Bel}(\phi) = \text{Bel}(\theta \wedge \phi).$$

Conversely suppose that $\text{Bel} : SL \rightarrow [0,1]$ satisfies (P1-2) and that for all θ, $\phi \in SL$ $\text{Bel}(\theta \mid \phi)$ is defined and satisfies $\text{Bel}(\theta \mid \phi)\text{Bel}(\phi) = \text{Bel}(\theta \wedge \phi)$. Then the fairness condition is satisfied.

(An alternative, and rather influential, justification for *Bel* as probability has been given by Cox [4] and Aczél [1, page 319].)

Notice that with this assumption the conditional constraints in K in question Q can be replaced by equivalent unconditional linear constraints provided we take as vaccous conditional constraints when the denominator is zero.

Thus we may assume that the set K of constraints in Q has the simpler form

$$\sum_{j=1}^{r} a_{ji}\,\text{Bel}\,(\theta_j) = b_i \qquad i = 1, ..., m$$

and restrict ourselves to the problem of assigning a value to $\text{Bel}(\theta)$.

The assumption that *Bel* should be a probability function is strongly supported by several arguments. Some useful and immediate consequences are:

Theorem 2 *For* $\theta, \phi \in SL$ *and* $\text{Bel} : SL \rightarrow [0,1]$ *a probability function,*

(a) $Bel(\neg\theta) = 1 - Bel(\theta)$

(b) *If* $\models \theta$ *then* $Bel(\neg\theta) = 0$

(c) *If* $\models (\theta \leftrightarrow \phi)$ *then* $Bel(\theta) = Bel(\phi)$

(d) $Bel(\theta \vee \phi) = Bel(\theta) + Bel(\phi) - Bel(\theta \wedge \phi)$

"

Further, with *Bel* being a probability we can find a simple and fruitful representation for it , as follows:

Suppose that $L = \{p_1, p_2,..., p_n\}$ and let $\alpha_1,..., \alpha_J$, where $J = 2^n$, run through the *atoms* of L, that is all conjunctions of the form

$$(\neg)^{\varepsilon_1} p_1 \wedge (\neg)^{\varepsilon_2} p_2 \wedge...\wedge(\neg)^{\varepsilon_n} p_n$$

where $\varepsilon_1,..., \varepsilon_n \in \{0,1\}$ and $(\neg)^0 p = p$, $(\neg)^1 p = \neg p$.

Then, by the disjunctive normal form theorem, for any $\phi \in$ SL there is a unique set

$$S_\phi \subseteq \{\alpha_1,..., \alpha_J \}$$

such that

$$\models \phi \leftrightarrow \bigvee S_\phi$$

Since $\models \neg(\alpha_i \wedge \alpha_j)$ for $i \neq j$, repeated use of (P2) gives

$$Bel(\phi) = Bel(\bigvee S_\phi) = \sum_{\alpha \in S_\phi} Bel(\alpha_i) \quad (even\ if\ S_\phi = \varnothing)$$

Also, since $\models \bigvee_{i=1}^{J} \alpha_i$, $1 = \sum_{i=1}^{J} Bel(\alpha_i)$ and of course the *Bel* $(\alpha_i) \geq 0$.

From this it follows that Bel is completely determined by its values on the atoms α_i, i.e. by the vector

$$< Bel(\alpha_1),...,Bel(\alpha_J)> \ \in \mathbf{D}^L = \{ \bar{x} \in \mathbf{R}^J \mid \bar{x} \geq 0, \sum_{i=1}^{J} x_i = 1\}$$

Conversely given $\bar{a} \in \mathbf{D}^L$ if we define a function $Bel': SL \rightarrow [0,1]$ by

$$Bel'(\phi) = \sum_{\alpha_i \in S_\phi} a_i$$

then Bel' is a probability function and furthermore

$$\bar{a} = \,< Bel'(\alpha_1),\ldots,Bel'(\alpha_J)>.$$

Summing up, what we have demonstrated here is that each probability function Bel (on SL) is determined by the point $<Bel(\alpha_1), \ldots, Bel(\alpha_J)> \in \mathbf{D}^L$ and conversely every point $\bar{a} \in \mathbf{D}^L$ determines a unique probability function Bel satisfying $<Bel(\alpha_1),\ldots,Bel(\alpha_J)> = \bar{a}$

This 1-1 correspondence between probability functions on L and points in $\mathbf{D}^L$ will be very useful in what follows. Indeed we shall frequently identify these notions, refering to a point in $\mathbf{D}^L$ as a probability function and conversely to a probability function as a point in $\mathbf{D}^L$.

Using this we can now reformulate question Q. Let K be as in the latest version of Q, say K is the set of constraints

$$\sum_{j=1}^{r} c_{ji} Bel(\theta_j) = b_i \qquad i = 1, \ldots, m$$

consistent with (P1-2), i.e. having a solution satisfying also (P1-2).

Now replace each $Bel(\theta_j)$ in K by $\sum_{\alpha_i \in S_{\theta_j}} Bel(\alpha_i)$ and, in matrix notation, let

$$< Bel(\alpha_1),\ldots,Bel(\alpha_J)> A_K = \bar{b}_K$$

be the resulting set of equations together with the equation

$$\sum_{i=1}^{J} Bel(\alpha_i) = 1$$

Then if *Bel* is a probability function satisfying K, the point $<Bel(\alpha_1),...,Bel(\alpha_J)>$ from $\mathbf{D}^L$ satisfies

$$\bar{x}A_K = \bar{b}_K, \qquad \bar{x} \geq 0$$

Conversely if $\bar{a} \in \mathbf{R}^J$ satisfies these equalities and inequalities, then $\bar{a} \in \mathbf{D}^L$ and the corresponding probability function *Bel* such that $Bel(\alpha_i) = a_i$, $i = 1, ..., J$ satisfies K.

Summing up then, the natural correspondence between probability functions and points in $\mathbf{D}^L$ gives a correspondence between solutions of K satisfying (P1-2) and points in

$$V^L(K) = \left\{ \bar{x} \in \mathbf{R}^J \middle| \bar{x}A_K = \bar{b}_K, \bar{x} \geq 0 \right\} \subseteq \mathbf{D}^L$$

The problem posed by Q, namely what value should we give to $Bel(\theta)$, for *Bel* satisfying K and (P1-2), now becomes, what value should we give to $\sum_{\alpha_i \in S_\theta} x_i$ given that $\bar{x}A_K = \bar{b}_K$, $\bar{x} \geq 0$. Of course in this reformulated version the connection with uncertain reasoning seems to have been 'abstracted away'. Nevertheless it will often be useful to consider the reformulated version in what follows.

Notice that the matrix A_K, as presented here, has size $J \times (m+1)$ where m is the original number of equations in K. In practice it is almost always the case that m is much less than J ($= 2^n$) so the system

$$\bar{x}A_K = \bar{b}_K, \quad \bar{x} \geq 0$$

rarely permits solving uniquely for $\bar{x}$ or indeed $\sum_{\alpha_i \in S_\theta} x_i$.

The fact that there may be many possible values for $\sum_{\alpha_i \in S_\theta} x_i$ for $\bar{x} \in V^L(K)$ means that we must seek further rational criteria for limiting this choice and providing answers to question Q.

One such source of criteria is common sense considerations and indeed that will be our main guiding light. [The criteria of consistency has already been implicitly applied in the sense that we are asking for a choice of a value from $\left\{ \sum_{\alpha_i \in S_\theta} x_i \,\middle|\, \bar{x} \in V^L(K) \right\}$ rather than simply $\left\{ \sum_{\alpha_i \in S_\theta} x_i \,\middle|\, \bar{x} \in \mathbb{D}^L \right\}$.] In view of the central role of 'common sense considerations' will play it seems that one might first, briefly, question why they are relevant at all in the context of question Q (as directed from situations such as example E). One such justification is, we believe, based on the common (but usually implicit) assumption that the elicited knowledge K is not simply a shadow or description of the Expert's knowledge but that, provided we have spent sufficient time collecting it,

K is (essentially) all the Expert's relevant knowledge

That is, that the Expert's knowledge *is* simply such a set of statements which given sufficient effort we can elicit. We shall refer to this as the Watt's Assumption.

If we make this assumption then our task of giving a value to $Bel(\theta)$ given K is exactly the task that the Expert himself carries out. Furthermore, since we might feel that his answer should be common sensical, applying common sense considerations to our assignment should still keep us within the realm of viable uncertain reasoning processes. [Despite its importance we shall avoid any philosophical or mathematical discussion of what we mean by common sense, or *why* we expect the Expert's assignments to be common sensical. Instead we hope the reader shares our view of what does or does not constitute common sense.]

Of course even if we do not make the Watts' Assumption we might feel it is unreasonable to violate common sense without strong reasons for so doing.

As an example of the sort of common sense considerations we have in mind consider the following example. Suppose that (under Watts' Assumption) the Expert actually knows nothing, i.e. $K = \emptyset$, where $L = \{ p_1, p_2 \}$ and he is asked to give a value to $Bel\,(p_1 \vee p_2)$. In this case any value between 0 and 1 is consistent with K (and Bel a probability function). However he might argue that since he knows nothing about p_1, p_2 each of the atoms (possible worlds!)

$$p_1 \wedge p_2, p_1 \wedge \neg p_2, \neg p_1 \wedge p_2, \neg p_1 \wedge \neg p_2 \quad (= \alpha_1, \alpha_2, \alpha_3, \alpha_4)$$

has, for him, equal status. Therefore since there are no grounds to distinguish them it would seem common sense to assign them all the same belief i.e. $\frac{1}{4}$ since we must have

$$Bel(p_1 \wedge p_2) + Bel(p_1 \wedge \neg p_2) + Bel(\neg p_1 \wedge p_2) + Bel(\neg p_1 \wedge \neg p_2) = 1$$

Hence, since

$$\models (p_1 \vee p_2) \leftrightarrow ((p_1 \wedge p_2) \vee (p_1 \wedge \neg p_2) \vee (\neg p_1 \wedge p_2)),$$

$$i.e. \ S_{p_1 \vee p_2} = \{\alpha_1, \alpha_2, \alpha_3\}, \ Bel(p_1 \vee p_2) = Bel(p_1 \wedge p_2) +$$

$$Bel(p_1 \wedge \neg p_2) + Bel(\neg p_1 \wedge p_2) = \frac{3}{4}.$$

As this example shows then there are principles of uncertain reasoning, based on common sense, which enable us to go beyond the assumption that Bel is a probability function and whose application may enable us to give an answer to question Q, or at least further restrict the possibilities. [As indicated earlier such common sense principles are not the only sort of considerations which might apply.]

It is our intention to investigate some such principles. However before we can do that we need to set up some apparatus in which to formulate them.

Recalling again our original example $\mathbb{E}$ it would have seemed irrational or inconsistent (in the natural language sense) if on one instance of question Q we had concluded from $\mathbb{K}_0$ that

B and F are never found together

and then, or a separate instance we had concluded from $\mathbb{K}_0$ that

The combination of B, F and not D is not uncommon

despite the fact that both these answers may be separately consistent with K_0.

From this example we see that, in the more general context of question Q, consistency requires that not only should the answers to various questions be separately consistent with K but furthermore they themselves should all be consistent with each other. That is, if given K we would assign value β_θ to $Bel\,(\theta)$ for $\theta \in SL$ then

$$K + \{\,Bel(\theta) = \beta_\theta \mid \theta \in SL\,\}$$

should be consistent i.e. there should be a probability function satisfying these. But clearly there could only be one such function since all its values are specified, indeed just its values on the $\alpha_1,...,\alpha_J$ would be enough.

Essentially then in assigning beliefs given K we are *picking* a probability function Bel to satisfy K, equivalently picking a point in $V^L(K)$. A belief assignment process which satisfies this extended notion of consistency will be called an Inference Process. Precisely let CL be the set of all finite consistent sets of linear constraints of the form

$$\sum_{j=1}^{r} a_j Bel(\theta_j) = b$$

where the $\theta_j \in SL$ and the b, $a_j \in \mathbb{R}$.

Definition 1 *An Inference Process on L is a function N such that for* $K \in CL$, *N(K) is a probability function, Bel, on SL satisfying K. That is N picks a solution to K or equivalently N picks a point in* $V^L(K)$.

Our main reason for introducing inference processes is that many common sense principles of uncertain reasoning can be formulated in terms of them. In what follows we shall consider some specific inference processes which have been proposed in the literature.

The general sorts of justification for these inference processes appear to fall under two or maybe three headings. In the first, as typified by CM^L,

CM_∞^L (see below), the point $N(K) \in V^L(K)$ is chosen to be as typical or average or as representative of the points in $V^L(K)$ as possible. In the second, $N(K)$ is chosen to be that point in $V^L(K)$ which 'contains as little information beyond K as possible', as typified by MD^L and ME^L.

Definition 2 *The Centre of Mass Inference Process for L, CM^L, is defined by $CM^L(K) = $ that point $< x_1,...,x_J >$ which is the centre of mass of $V^L(K)$, assuming uniform density, i.e.*

$$CM^L(K)(\alpha_i) = \frac{\int V^L(K)^{x_i} dV}{\int V^L(K) dV}$$

where the integrals are taken over the relative dimension of $V^L(K)$ using the same basis.

CM^L, may be, in part, justified by the so called *Principle of Indifference* (or Laplace's principle) by which we mean that given some facts all the possible worlds consistent with the facts should be viewed as equally likely. In this case then each point in $V^L(K)$ should be given equal weight and the choice of the centre of mass (with uniform density) just corresponds to picking the average or most representative of the points in $V^L(K)$.

A criticism of CM^L, is that if we view this process also as a function of the finite language L then it is not language invariant.

Definition 3 *Suppose we have a family of inference processes N^L, one for each finite language L. Then this family is said to be language invariant if whenever $L_1 \subseteq L_2$ (so $SL_1 \subseteq SL_2$ and $CL_1 \subseteq CL_2$) and $K \in CL_1$ then $N^{L_2}(K)$ agrees with $N^{L_1}(K)$ on SL_1.*

The CM^L, are not language invariant, adding additional propositional variables to the language will in general alter the belief given to θ on the basis of K even though the new variables do not explicitly appear in θ or K. One way to correct this variability as the size of the overlying language changes is to consider the limiting inference process we obtain by saturating the context with new propositional variables. This yields the language invariant inference process CM_∞^L , where

Definition 4 CM_∞^L *is the inference process on L which on argument K picks that* $\bar{x} \in V^L(K)$ *at which the function* $\sum_{i \notin I^L(K)} \log(x_i)$ *is maximal where*

$$I^L(K) = \left\{ i \mid \forall \bar{x} \in V^L(K), x_i = 0 \right\}$$

(so $x_i = 0$ *) for* $i \in I^L(K)$*).*

An alternative class of inference processes arise from picking the point in $V^L(K)$ with the 'minimum information content'. One example of this is the following

Definition 5 *The Minimum Distance inference process,* MD^L*, is defined by*

$$MD^L(K) = \text{the nearest (usual metric) point in } V^L(K)$$
$$\text{to} < \frac{1}{J}, ..., \frac{1}{J} > \in \mathbb{R}^J$$
$$= \text{that } \bar{x} \in V^L(K) \text{for which } \sum_{i=1}^J (x_i - \frac{1}{J})^2$$
$$\text{is minimal.}$$

A justification for picking this point is that since the point $< \frac{1}{J}, \frac{1}{J}, ..., \frac{1}{J} >$ represents the least informative belief function (in that it gives no preferences amongst the atoms, or possible worlds) then in order to pick the least informative point in $V^L(K)$ equivalently the point that contains as little information beyond K as possible, we should pick that point in $V^L(K)$ closest to the point representing total ignorance, that is $< \frac{1}{J}, \frac{1}{J}, ..., \frac{1}{J} >$. MD^L is language invariant.

A second inference process based on information (or lack of it) is

Definition 6 *The Maximum Entropy Inference Process ME^L, is defined by*

$$ME^L(K) = \quad \text{that } \bar{x} \in V^L(K) \text{ for which the entropy,}$$
$$\text{i.e. Shannon measure of uncertainty,}$$
$$-\sum_{i=1}^{J} x_i \log(x_i), \text{ is maximal}$$

In this process then we choose the point which has the greatest measure of uncertainty, or equivalently the least information, according to Shannon's measure. Again then we are picking that point in $V^L(K)$ which goes as little beyond the information given by the constraints K as possible.

IV. PRINCIPLES OF UNCERTAIN REASONING

Above we introduced a selection of the more popular inference processes which have been proposed. This raises the question of why to prefer one such process over any other. In this chapter we shall consider this question by presenting a number of properties, or as we shall call them, principles, which it might be deemed desirable that an inference process, N, should satisfy.

For the most part these principles could be said to be based on common sense or rationality or 'consistency' in the natural language sense of the word. A justification for assuming that adherence to common sense is a desirable property of an inference process comes from the Watts' Assumption quoted earlier. For if K genuinely does represent all the Expert's knowledge then any conclusion the expert draws from K should be the result of applying what, by concensus, we consider correct reasoning i.e. by common sense.

So our plan now is to present a list of such principles. We limit ourselves to the framework where *Bel* is a probability function, although the same criteria could be applied to inference processes for Shafer-Dempster belief functions, possibility functions etc. In what follows N stands for an inference process on L. Here L is to be thought of as variable. If we wish to consider a principle for a particular language L we shall insert 'for L'.

V. EQUIVALENCE PRINCIPLE

If $K_1, K_2 \in CL$ are equivalent in the sense that $V^L(K_1) = V^L(K_2)$ then $N(K_1) = N(K_2)$.

That is, if K_1, K_2 put exactly the same constraints on a probability function Bel (in that Bel satisfies one iff it satisfies the other) then the principle asserts that $N(K_1)$, $N(K_2)$ should be equal. The principle implies then that N is a function of the convex sets $V^L(K)$ rather than the $K \in CL$ from which they are derived.

This principle may be justified directly from the Watts' Assumption by arguing that it is implicit in this assumption that the knowledge *is* the constraints it puts on Bel rather than the particular way in which the constraints are expressed.

Since all of the processes CM^L, CM_∞^L, MD^L, ME^L are defined in terms of $V^L(K)$, they all satisfy the principle of Equivalence. Henceforth in this chapter when presenting other principles we shall assume that N satisfies the principle of equivalence.

VI. PRINCIPLE OF IRRELEVANT INFORMATION

Suppose that $K_1, K_2 \in CL, \theta \in SL$ but that no propositional variable appearing in θ or in any sentence in K_1 also appears in a sentence in K_2. Then $N(K_1 + K_2)(\theta) = N(K_1)(\theta)$. [Here $K_1 + K_2$ is the union of these two sets of constraints.]

The justification for this principle is that the knowledge provided by K_2 is irrelevant to K_1 and θ since it concerns a completely separate set of propositional variables (i.e. features) and hence the belief given to θ on the basis of $K_1 + K_2$ should be the same as on the basis of K_1 alone.

We note here that for such separate $K_1, K_2 \in CL, K_1 + K_2$ is consistent, and hence in CL. To see this suppose that Bel_1, Bel_2 are probability functions on SL satisfying K_1, K_2 respectively, and suppose that $K_1 \in CL_1, K_2 \in CL_2$ where $L_1 \cup L_2 = L, L_1 \cap L_2 = \emptyset$. Let $\alpha_1^1, \dots, \alpha_{J_1}^1$

and $\alpha_1^2,...,\alpha_{J_2}^2$ be the atoms of L_1, L_2 respectively so that (up to logical equivalence) the atoms of L are $\alpha_i^1 \wedge \alpha_j^2$ for $i = 1,..., J_1, j = 1,..., J_2$. Define a probability function Bel on SL by

$$Bel(\alpha_i^1 \wedge \alpha_j^2) = Bel_1(\alpha_i^1)Bel_2(\alpha_j^2).$$

Then Bel agrees with Bel_1 on SL_1, since

$$Bel(\alpha_i^1) = \sum_{j=1}^{J_2} Bel(\alpha_i^1 \wedge \alpha_j^2) = Bel_1(\alpha_i^1)\sum_{j=1}^{J_2} Bel_2(\alpha_j^2) = Bel_1(\alpha_i^1),$$

and with Bel_2 on SL_2, so Bel satisfies both K_1 and K_2.

The Maximum Entropy Inference Process, ME, satisfies the Principle of Irrelevant Information. However none of CM^L, MD^L, CM_∞^L satisfy this principle. This is clear for CM^L, since, by taking $K_2 = \emptyset$, we see that this principle would imply language invariance.

VII. CONTINUITY

The essence of this principle, or desideratum, is that, for $\theta \in SL$, a microscopic change in the knowledge K should not cause a macroscopic change in the value $N(K)(\theta)$ of the belief assigned to θ on the basis of K. The justification for this is that it is unreasonable to expect that the knowledge K of the Expert is not subject to minor change fluctuations even if he receives no new knowledge. Nevertheless it would seem inconsistent, in the natural language sense, if this caused appreciable changes in his beliefs. Thus common sense, (and the subsequent commitments they engender) or at least our common understanding of the function of belief as condoning actions, requires that the values $N(K)(\theta)$ display a certain robustness in the face of minor fluctuations in the knowledge K.

In formulating this principle however we are faced with the problem of finding a metric on CL which captures the idea that nearness in the metric corresponds to nearness in knowledge content. An immediate response might well be to use some standard metric on the matrices $\begin{pmatrix} \bar{b}_K \\ A_K \end{pmatrix}$, for example

$$\| A - B \| = \max \mid a_{ij} - b_{ij} \mid$$

where $A = (a_{ij})$, $B = (b_{ij})$. Indeed this appears on the face of it to be just what is required since it would seem that a 'minor fluctuation' of K would correspond exactly to a minor fluctuation in the coefficients α_{ij}, β_i appearing in K and hence to a matrix close to A_K in this metric.

Unfortunately this metric does not capture the idea of distance between knowledge content as the following example shows. Suppose $L = \{p\}$ and

$$K = \{0Bel\ (p) = 0\}, K_\varepsilon = \{\varepsilon Bel\ (p) = 0\}$$
$$K'_\varepsilon = \{\varepsilon Bel\ (\neg p\ ) = 0\}, \quad K''_\varepsilon = \{\varepsilon Bel\ (p\ ) = \varepsilon Bel\ (\neg p\ )\}$$

Then

$$\begin{pmatrix} 0 \\ A_{K_\varepsilon} \end{pmatrix}, \begin{pmatrix} 0 \\ A_{K'_\varepsilon} \end{pmatrix} \begin{pmatrix} 0 \\ A_{K''_\varepsilon} \end{pmatrix} \rightarrow \begin{pmatrix} 0 \\ A_K \end{pmatrix} \quad as \ \varepsilon \rightarrow 0$$

in the above metric. However it is clear that in terms of the knowledge they express $K_\varepsilon, K'_\varepsilon, K''_\varepsilon$ for $\varepsilon > 0$, are about as distant from each other as one could imagine. Of course one could argue that this is a rather transparent and unnatural example. However there are much more subtle (and natural) examples, as the following, due to P. Courtney, shows.

Suppose a certain disease d has a symptom s and possible complication c (so $L = \{d,s,c\}$. A Medical Expert gives the following as his knowledge about their relationships:

1. $Bel\ (d \wedge s) = 0.75\ Bel\ (s)$ 4. $Bel\ (d \wedge \neg s) = 0.25\ Bel\ (\neg s)$

2. $Bel\ (\neg c \wedge d \wedge s) = 0.15\ Bel\ (s)$ 5. $Bel\ (\neg c \wedge d \wedge \neg s) = 0.6\ Bel\ (d \wedge \neg s)$

3. $Bel\ (c \wedge d \wedge s) = 0.8\ Bel(d \wedge s)$ 6. $Bel\ (c \wedge d \wedge \neg s) = 0.1\ Bel\ (\neg s)$

These are certainly consistent, for example take $Bel\ (s) = 1$, $Bel\ (c \wedge d \wedge s)$ $= 0.6$, $Bel\ (\neg\, c \wedge d \wedge s) = 0.15$, $Bel\ (c \wedge \neg\, d \wedge s) = 0.25$ and all the other atoms zero.

However since the equations 4, 5, 6, force

$$
\begin{aligned}
Bel\ (d \wedge \neg s) \quad &= \quad Bel\ (c \wedge d \wedge \neg s) + Bel\ (\neg\, c \wedge d \wedge \neg s) \\
&= \quad (0.6 + \tfrac{0.1}{0.25})Bel(d \wedge \neg s)
\end{aligned}
$$

any change in any one coefficient in 4,5,6 forces $Bel(d \wedge \neg s) = Bel(\neg s) = 0$. Similarly any change in any one coefficient in 1,2,3 forces $Bel(s) = 0$.

The conclusion that such examples forces upon us is that closeness of $\begin{pmatrix} \bar{b}_K \\ A_K \end{pmatrix}, \begin{pmatrix} \bar{b}_{K'} \\ A_{K'} \end{pmatrix}$ (in the above metric) does not correspond to closeness in knowledge content (although we must admit it is questionable whether or not the above intuitions about the common sense concerning continuity were not in fact based on closeness in the former sense!). What is required here is a metric which better corresponds to distance between knowledge contents. One natural criterion for K, K' to be close to each other is that every solution, Bel, of K is close to a solution of K' and vice-versa. When expressed in terms of the convex sets $V^L(K)$ this is just the well known Blaschke metric, Δ, defined, for convex subsets C, D of $\mathbb{D}^L$ by

$$
\Delta(C,D) = \mathit{Inf}\,\{\ \delta \big| \forall \bar{x} \in C \exists \bar{y} \in D, \big|\,\bar{x} - \bar{y}\,\big| \le \delta \wedge \forall \bar{y} \in D \exists \bar{x} \in C, \big|\,\bar{x} - \bar{y}\,\big| \le \delta\ \}
$$

where $|\ \bar{x} - \bar{y}\,|$ is the usual Euclidean distance between the points $\bar{x}$ and $\bar{y}$.

We can now state the continuity requirement on an inference process N as the requirement that N, as a function of the convex sets $V^L(K)$ is continuous with respect to this Blaschke metric, i.e. If $\theta \in SL, K, K_i \in CL$ for $i \in \mathbb{N}$ and $\lim_{i \to \infty} \Delta(V^L(K), V^L(K_i)) = 0$ then $\lim_{i \to \infty} N(K_i)(\theta) = N(K)$ (θ).

Notice that it is enough here to restrict θ to being an atom. The following theorem, due to I. Maung [11], shows that ME, CM_∞, MD are continuous.

Theorem 3   *If the process N on L satisfies $N(K) =$ that $\bar{x} \in V^L(K)$*
at which $F(\bar{x})$ is maximal for some (strictly) convex
function $F : \mathbb{D}^L \to \mathbb{R}$ then N is continuous.

The inference process CM^L does not satisfy continuity for this
metric. For example for $L = \{\, p, q\,\}$ and K_ε ($\varepsilon \geq 0$) being

$$(1 - \varepsilon)\, Bel\, (\neg\, p \wedge q\,) + Bel\, (\neg\, p \wedge \neg\, q\,) = 1 - \varepsilon$$
$$(1 - \varepsilon)\, Bel\, (p) - \varepsilon\, Bel\, (\neg\, p \wedge \neg\, q\,) = 0$$

it is the case that $CM^L(K_0)(\neg p \wedge q) = \frac{1}{2}$ whilst $CM^L(K_\varepsilon\}) = \frac{1}{3}$ for $\varepsilon > 0$
despite the fact that $V^L(K_\varepsilon)$ tends to $V^L(K_0)$ in the Blaschke topology as ε
tends to zero. The reason for this discontinuity here is that for $\varepsilon > 0$, $V^L(K_\varepsilon)$
is an isoceles triangle with base on $x_3 = 0$ and apex on $x_3 = 1$ so its centre
of mass satisfies $x_3 = \frac{1}{3}$. As ε tends to zero however the sides fold together
yielding in the limit a line with centre of mass satisfying $x_3 = \frac{1}{2}$.

Returning again to the earlier, flawed, idea that continuity of an
inference process N at K should mean that if

$$\left\| \begin{pmatrix} \bar{b}_K \\ A_K \end{pmatrix} - \begin{pmatrix} \bar{b}_{K'} \\ A_{K'} \end{pmatrix} \right\|$$

is small then $N(K)$ is close to $N(K')$, it is now natural to ask under what
conditions

$$\left\| \begin{pmatrix} \bar{b}_K \\ A_K \end{pmatrix} - \begin{pmatrix} \bar{b}_{K'} \\ A_{K'} \end{pmatrix} \right\|$$

being small forces $V^L(K)$, $V^L(K')$ to be close in the Blaschke topology.

In fact there are two situations in which this can fail. The first is
when $I^L(K) \neq I^L(K')$, where $I^L(K) = \{\, i \mid \forall \bar{x} \in V^L(K), x_i = 0 \,\}$, the second

when $\hat{A}_K, \hat{A}_{K'}$ have different ranks where $\hat{A}_K$ is the result of deleting the i-th row from A_K for each $i \in F(K)$. Indeed these are the reasons continuity fails in the two examples given earlier. The following theorem, due to P. Courtney [3], shows that if these situations do not occur then closeness of the coefficients in K' to those in K does indeed force $V^L(K)$, $V^L(K')$ to be close in the Blaschke sense.

Theorem 4 *Let $K \in CL$ and $\varepsilon > 0$. Then there is $\delta > 0$ such that if $K' \in CL$, $A_{K'}$ has the same number of rows as A_K and*

> (i) $Rank(\hat{A}_K) = Rank(\hat{A}_{K'})$,
>
> (ii) $I^L(K) = I^L(K')$,
>
> (iii) $\left\| \begin{pmatrix} \bar{b}_K \\ A_K \end{pmatrix} - \begin{pmatrix} \bar{b}_{K'} \\ A_{K'} \end{pmatrix} \right\| < \delta$,
>
> *then $\Delta(V^L(K), V^L(K')) < \varepsilon$.*

We finally remark, omitting the proof, that if $K, K_n \in CL$ are such that

$$\left\| \begin{pmatrix} \bar{b}_K \\ A_K \end{pmatrix} - \begin{pmatrix} \bar{b}_{K_n} \\ A_{K_n} \end{pmatrix} \right\| \to 0 \text{ as } n \to \infty \text{ and } Rank(\hat{A}_K) = Rank(\hat{A}_{K_n}), I^L(K) =$$

$I^L(K_n)$ for all n then $\lim_{n \to \infty} CM^L(K_n) = CM^L(K)$, so CM^L is continuous in this weaker sense.

Nevertheless in this paper we shall take continuity for an inference process to mean continuity for the Blaschke topology.

VIII. OPEN MINDEDNESS PRINCIPLE

If $K \in CL, \theta \in SL$ and $K + Bel(\theta) \neq 0$ is consistent then $N(K)(\theta) \neq 0$.

The justification for this principle is that if it is consistent with K that θ is possible (i.e. $Bel(\theta) \neq 0$) then the inference process N should not dismiss, or classify, θ as impossible since this would amount to introducing assumptions beyond those in K. This principle highlights the very special nature of categorical beliefs (i.e. with values 0,1) as compared with beliefs in the range (0,1). (An explanation of this is, perhaps, that beliefs in the range (0,1) are perceived as having no absolute meaning).

It is easy to convince oneself by a geometric argument that MD^L does not satisfy this principle whilst CM^L does. ME^L and CM^L_∞ also satisfy this principle, essentially because the derivatives of $-x \log (x)$ and $\log (x)$ tend to infinity as x tends to zero.

IX. RENAMING PRINCIPLE

Suppose $K_1, K_2 \in CL$,

$$K_1 = \{ \sum_{j=1}^{J} a_{ji} Bel(\gamma_j) = b_i \mid i = 1,...,m \},$$

$$K_2 = \{ \sum_{j=1}^{J} a_{ji} Bel(\delta_j) = b_i \mid i = 1,...,m \},$$

where $\gamma_1,...,\gamma_J$, $\delta_1,...,\delta_J$ are permutations of the atoms $\alpha_1,...,\alpha_J$ of L. Then $N(K_1)(\gamma_j) = N(K_2)(\delta_j)$

The justification for this principle is that the atoms of SL all share the same status of being simply possible worlds and so the particular ordering $\alpha_1,...,\alpha_J$ of these atoms which we chose should not be significant. Whilst this principle is comparatively easy to accept in this form (which is why we presented it thus) it gains much more strength when used in conjuction with language invariance and the principle of irrelevant information.

Theorem 5 *Suppose*

$$K_1 = \{ \sum_{j=1}^{r} a_{ji} Bel(\theta_j) = b_i \mid i = 1,\ldots,m \} \in CL_1,$$

$$K_2 = \{ \sum_{j=1}^{r} a_{ji} Bel(\phi_i) = b_i \mid i = 1,\ldots,m \} \in CL_2,$$

and σ is a bijection between the At^{L_1} and At^{L_2} such that $\sigma'' S_{\theta_j}^{L_1} = S_{\phi_j}^{L_2}$ for $j = 1,\ldots,r$ and $\sigma'' S_{\theta}^{L_1} = \sigma'' S_{\phi}^{L_2}$. Then for an inference process N satisfying language invariance and the principles of renaming and irrelevant information,

$$N(K_1)(\theta) = N(K_2)(\phi).$$

Notice that in the special case when we have a $1 - 1$ onto map η from L_1 to L_2 and we extend η to SL_1, CL_1 etc and $\eta(K_1) = K_2$, $\eta(\theta) = \phi$ the theorem shows that for N satisfying language invariance and the principles of renaming and irrelevant information,

$$N(K_1)(\theta) = N(K_2)(\phi)$$

In this case the conclusion seems so especially reasonable that it deserves the title *Weak Renaming Principle*.

We can reformulate the renaming principle in terms of the convex sets $V^L(K)$ as follows. For σ a permutation of $1,\ldots, J$ extend σ to $\mathbb{R}^J$ by

$$\sigma < x_1,\ldots,x_J > = < x_{\sigma(1)},\ldots,x_{\sigma(J)} >.$$

Then N satisfying renaming principle is equivalent to N satisfying

$$\sigma N(V^L(K)) = N(\sigma V^L(K))$$

for $K \in CL$, σ a permutation of $1,...,J$ where here we are directly treating N as a choice function on the sets $V^L(K)$.

The renaming principle holds for the inference processes CM^L, CM_∞^L, MD^L, ME^L since they are all easily seen to have this latter property that $\sigma N(V_K^L) = N(\sigma V_K^L)$. This is simply because these processes do give all the variables x_i (i.e. $Bel(\alpha_i)$) equal status. Nevertheless this principle seems not beyond criticism. Certainly it is hard to object to it in the case where say $\gamma_1,...,\gamma_J$ is just the result of transposing all occurences of, say the propositional variables p_s, p_t in the atoms $\delta_1,...,\delta_J$ (i.e. $p_1^{\varepsilon_1} \wedge...\wedge p_s^{\varepsilon_s} \wedge...$ $\wedge p_t^{\varepsilon_t} \wedge...\wedge p_n^{\varepsilon_n}$ would become $p_1^{\varepsilon_1} \wedge...\wedge p_s^{\varepsilon_t} \wedge...\wedge p_t^{\varepsilon_s} \wedge...\wedge p_n^{\varepsilon_n}$) or of transposing $p_s, \neg p_s$ (i.e. $p_1^{\varepsilon_1} \wedge...\wedge p_s^{\varepsilon_s} \wedge...\wedge p_n^{\varepsilon_n}$ would become $p_1^{\varepsilon_1} \wedge..\wedge p_s^{1-\varepsilon_s}$ $\wedge..\wedge p_n^{\varepsilon_n}$) since these look like genuine 'renamings'. However arbitrary permutations are perhaps rather stretching the notion of renaming. For example in such a way for $L = \{p_1, p_2, p_3\}$, the constraint $Bel(p_1 \wedge p_2) = b$ (equivalently $Bel(p_1 \wedge p_2 \wedge p_3) + Bel(p_1 \wedge p_2 \wedge \neg p_3) = b$) could be 'renamed' as $Bel(\neg p_1 \wedge p_2 \wedge p_3) + Bel(p_1 \wedge \neg p_2 \wedge p_3) = b$ and in this case one might feel that renaming has changed the content of the constraint. [For an interesting discussion on this point see D. Miller [10]. If an inference process N satisfies renaming then it also satisfies the following:

X. PRINCIPLE OF INDIFFERENCE

If $K = \{Bel(\theta) = 1\}$ where $\theta \in SL$ then $N^L(K)(\alpha_i) = |S_\theta|^{-1}$ for $\alpha_i \in S_\theta$ (and necessarily zero otherwise). In other words all worlds consistent with K should get equal belief, or probability.

To see this notice that K is equivalent to the single constraint

$$\sum_{\alpha_i \in S_\theta} Bel(\alpha_i) = 1$$

Now if α_k, $\alpha_j \in S\theta$ and we consider the permutation of the atoms which simply transposes α_k, α_j then K will not alter so by the renaming principle

$$N^L(K)(\alpha_k) = N^L(K)(\alpha_j).$$

The result now follows since $N^L(K)$ must satisfy K, i.e.

$$\sum_{\alpha_i \in S_\theta} N^L(K)(\alpha_i) = 1$$

We shall refer to this as the Finite Principle of Indifference (since there are only finitely many possible worlds) to distinguish it from the discredited Infinite Principle of Indifference considered in the previous chapter.

Notice that the set of constraints

$$Bel(\theta_i) = 1, \quad i = 1,...,m \quad Bel(\neg \phi_j) = 0, \quad j = 1,..., q$$

is equivalent to the single constraint

$$Bel(\bigwedge_{i=1}^{m} \theta_i \wedge \bigwedge_{j=1}^{q} \neg\phi_j) = 1$$

so that the above principle can be applied whenever our constraint set K consists solely of categorical constraints.

XI. OBSTINACY PRINCIPLE

Suppose $K_1, K_2 \in CL$ and $N(K_1)$ satisfies K_2. Then $N(K_1 + K_2) = N(K_1)$.

The justification for this principle is that since $N(K_1)$ satisfies K_2, K_2 would (according to the inference process N) be believed on the basis of

K_1 and so adding to K_2 to K_1 provides no more information, or alternatively gives no grounds for changing beliefs.

It is easy to convince oneself with a geometric argument that except in the case $|L| = 1$, CM^L does not satisfy obstinacy. However ME^L, CM_∞^L, MD^L all do satisfy this principle. Indeed this is a consequence of the general fact that if an inference process N^L can be defined by

$$N^L(K) = \text{ the unique maximum point } \bar{x} \in V^L(K) \text{ of } F^L$$

where F^L is a function from $\mathbb{D}^L$ into a linearly ordered set, then N^L is obstinate (for L). This follows since if $\bar{x}$ is the maximum point of F^L in $V^L(K_1)$ and $\bar{x} \in V^L(K_2)$ then $\bar{x} \in V^L(K_1) \cap V^L(K_2) = V^L(K_2) = V^L(K_1 + K_2)$ and, because $V^L(K_1 + K_2) \subseteq V^L(K_1)$, $\bar{x}$ must also be the maximum point of F^L on this set.

From this example it might be hoped that every obstinate inference can be represented in this way for some such F^L. However by considering a two dimensional spiral examples of obstinate inference processes can be found which are not of this form.

XII. RELATIVISATION PRINCIPLE

Suppose $0 < c < 1$ and $K_1, K_2 \in CL$ are respectively

$$Bel(\phi) = c, \quad \sum_{j=1}^{r} a_{ji} Bel(\theta_i \mid \phi) = b_i \quad i = 1,...,m,$$

$$K_1 + \sum_{j=1}^{q} e_{ji} Bel(\psi_i \mid \neg\phi) = f_i \quad i = 1,...,s.$$

Then for $\theta \in SL, N(K_1)(\theta \mid \phi) = N(K_2)(\theta \mid \phi)$.

Here conditional beliefs are to be interpreted, of course, as conditional probabilities. The justification for this principle is that K_1, K_2 both give the same belief to ϕ and given ϕ they both express exactly the same knowledge (constraints) - since given ϕ the additional knowledge in K_2 becomes vaccous. Hence the conditional belief given to θ given ϕ by N on the basis of K_1 should be the same as that given on the basis of K_2.

This principle fails for CM^L, CM_∞^L, MD^L but holds for ME^L.

XIII. PRINCIPLE OF INDEPENDENCE

In the special case of $L = \{p_1, p_2, p_3\}$ and $K = \{ Bel\ (p_1) = a, Bel\ (p_2 \mid p_1) = b, (Bel\ (p_3 \mid p_1) = c\}$, where $a > 0$, $N^L\ (K)\ (p_2 \wedge p_3 \mid p_1) = bc$.

Again here conditional beliefs are identified with conditional probabilities. The justification for this principle is that since beliefs are probabilities and in K there is no material connection between p_2 and p_3 given p_1, so p_2, p_3 should be treated as statistically independent given p_1, i.e.

$$Bel(p_2 \wedge p_3 \mid p_1) = Bel(p_2 \mid p_1).\ Bel(p_3 \mid p_1),$$

which yields the conclusion of the principle. (Of course whether or not statistical independence exactly captures ones intuitive idea of independence might perhaps not be universally accepted.

ME^L satisfies this principle, as can easily be checked, whilst CM^L, CM_∞^L, MD^L do not, yielding for this K in the case $a = 1, b = c = \frac{1}{n}, n \geq 2$ the answers $\frac{1}{2n}$, $\frac{8 - 3n + \sqrt{9n^2 - 32n + 32}}{8n}$, 0 respectively. Interestingly however in the case $K = \{ Bel\ (p_1) = a, Bel\ (p_2 \mid p_1) = b \}$ we find $ME\ (K)\ (p_2) = ab + \frac{1}{2}(1 - a)$ not the expected (?) answer b.

In so far as the principles introduced so far could be said to be principles of common sense or rationality they clearly provide an argument in favour of ME^L, the Maximum Entropy Inference process, as against CM^L, CM_∞^L, MD^L, since ME^L is the only one of these processes to satisfy all these principles. In fact they provide an argument favouring ME^L over any other inference process as the following result due to the authors [12] shows.

Theorem 6 *ME is the only inference process satisfying continuity and the principles of equivalence, irrelevant information, open mindedness, renaming, obstinacy, relativisation and independence.*

Since common sense, or rationality, was used to justify belief as probability one could claim that *ME* is the only inference process consistent with common sense. We should be aware however that in coming to this conclusion we have made the assumption that the knowledge statements K themselves constitute the knowledge. Without this assumption the very relevance of common sense principles is questionable.

We now go on to consider the two other principles, or at least desiderata.

XIV. ATOMICITY PRINCIPLE

Let $\theta \in SL_2$ be neither a contradiction nor a tautology, $K \in CL_1, \phi \in SL_1, L_1 \cap L_2 = \emptyset$ and $L_1, L_2 \subseteq L$. Let ϕ^θ etc. be the result of replacing a particular propositional variable $p \in L_1$ everywhere in ϕ by θ. Then $N^L(K)(\phi) = N^L(K^\theta)(\phi^\theta)$.

This principle might be justified by arguing that in practice the things we denote by propositional variables are determined not by some intrinsic property that they possess but simply because we choose to stop the depth of our analysis at this point. Presumably this reflects our belief that analysing them more deeply would not cause our answers to alter.

Unfortunately this likely looking principle is inconsistent with belief as probability! To see that suppose $K = \emptyset, L = \{p_1, p_2, p_3\}$. Taking $\phi = p_1$ and $\theta = p_2 \wedge p_3, p_2 \wedge \neg p_2, p_2, \neg p_2$ we see that $N^L(\phi)$ would have to give each of these the same value and, combined with $N^L(\phi)(p_2) + N^L(\phi)(\neg p_2) = 1$ gives a contradiction.

Nevertheless this principle is consistent, and indeed is satisfied by *ME*, if we add the requirement that $(Bel\ (p) = a) \in K$ for some a (although it is hard to see that the intuition, which previously justified an untenable principle, is in any way altered by this insertion).

The principles we have introduced so far were motivated by considerations of common sense or rationality. However there are certainly other sources which might inspire principles or at least desiderata. One such source is considerations of computational feasibility although we shall not investigate it in this paper.

Another is the requirement of accuracy. After all in practice accuracy is the main priority - all the common sense in the worlds is of no value if it produces woefully incorrect answers. This, of course, raises the question of what 'correct' means, or could mean, in the context of uncertain reasoning, especially as regards the relationship of the knowledge statements K to the real world. Interesting as this question is however we shall resist the temptation and, for this paper at least, we shall content ourselves with considering one, rather strong, notion of correctness of a solution, namely, of being *calibrated*.

XV. CALIBRATION

Let $K \in CL$. Then we shall say that a solution Bel_0 to K is *calibrated* if for each $\theta \in SL$ and $w \in V^L(K)$,

$$Bel_0(\theta) = \frac{1}{q}\sum_{i-1}^{q} w(\theta_i)$$

where $\{ S_{\theta_1},...,S_{\theta_q} \} = \{ R \subseteq \{ \alpha_1,...,\alpha_J \} \mid Bel_0(VR) = Bel_0(\theta) \}$.

This version of calibration was introduced by P. Lehner [9] (there called expected calibration) developing ideas of P. Horwich [6]. To see a relationship between calibration and the, presumably, desirable property of accuracy consider the following example. An expert maintains that some particular outcome is almost certain to occur but in the event it does not occur. Well, he could be excused, after all he did not actually assert that the outcome was certain so he was acknowledging that there was an outside chance that the outcome might fail to materialise. [For example he might, quite reasonably, inform anyone who asked that their lottery ticket was almost certainly not going to win. Nevertheless we would not expect all his answers to be vindicated since *some one* is going to win.] But now suppose that this happened repeatedly, that the Expert claimed a particular outcome was almost certain but, in the event, it did not occur. In that case we would start to seriously doubt the accuracy of the Experts beliefs, indeed, more importantly, the Expert himself should have doubts about his beliefs.

Assuming then that the Expert has well thought out beliefs which he is not inclined to change, his beliefs should be such that, in his view, they do not seriously risk him committing a high proportion of (distinct) errors. To take a slightly more precise example suppose the expert answers a large questionnaire concerning his beliefs in certain statements. Then of those statements about which he expresses belief b we might expect that, in the event, a proportion b, or there abouts, of them would in fact turn out to be true. Certainly a proportion here differ significantly from b should cause the expert to re-examine his beliefs and the mechanism by which he arrived at them.

Now suppose that the expert has given a constraint set K and we make the Watt's Assumption that K *is* exactly his knowledge and hence that his further beliefs are made on the basis of 'knowledge' K alone. Then in infering beliefs from K the expert would, presumably, not wish to risk being inaccurate in the above sense. Indeed if this was his top priority, he might prefer not to seek higher principles on which to base his particular choice of a belief function satisfying K but rather, if possible, to assign beliefs in such a way that he runs no risk of consistently erring in this fashion no matter which choice of belief function is 'correct'. That is, in asserting K, he is asserting that, in his opinion, the correct answer satisfies K, but rather than go further and make a definite and possibly inaccurate, choice of 'correct', he prefers to assign beliefs in such a way that he runs no risk of consistently erring in the above fashion, no matter what the 'correct' belief function satisfying K is.

Formally then, if $\theta_1,...,\theta_q$ are exactly the distinct outcomes to which the Expert gives belief b on the basis of some general knowledge K then he should ideally prefer that for any probability function w satisfying K the expected proportion of $\theta_1,...,\theta_q$ which occur, according to w, is b, i.e.

$$b = \frac{1}{q}\sum_{i=1}^{q} w(\theta_i)$$

Identifying 'distinct outcomes' here with a maximal set of inequivalent outcomes then provides the justification for choosing, if possible, a calibrated solution.

Unfortunately calibrated solutions do not always exist for $K \in CL$ although as we shall see, if a calibrated solution does exist then it is unique.

Our aim now is to give a characterisation of the calibrated solution, when it exists, due to P. Lehner [9], and then to connect calibration with two principles mentioned earlier.

The following theorem and discussion is due to P.Lehner [9].

Theorem 7 *Let $K \in CL$ and $Bel_0 \in V^L(K)$. Then Bel_0 is calibrated iff Bel_0 is calibrated with respect to atoms, that is, for $1 \le i \le J$ and $w \in V^L(K)$,*

$$Bel_0(\alpha_i) = \frac{1}{|T_i|}\sum_{\alpha_j \in T_i} w(\alpha_j)$$

where $T_i = \{ \alpha_j | Bel_0(\alpha_i) = Bel_0(\alpha_j) \}$.

Discussion

Suppose $K \in CL$ has a calibrated solution, Bel_0, say, and let $T_1,...,T_s$ be as in the above theorem. Then since for $1 \le i \le s$ and all $w \in V^L(K)$,

$$w(\bigvee T_i) = \sum_{\alpha \in T_i} w(\alpha) = |T_i| Bel_0(\alpha_i)$$

we see that, the $\bigvee T_i$ are exclusive and exhaustive and the constraints

$$Bel(\bigvee T_i) = |T_i| Bel_0(\alpha_i) \qquad i = 1,...,s$$

are consequences of K (indeed they are formally derivable from K in a suitable proof theory).

Furthermore every calibrated solution arises in this way. For suppose $\psi_1,...,\psi_s$ are exclusive and exhaustive, $K \in CL$, and the constraints

$$Bel(\psi_i) = c_i \qquad i = 1,...,s$$

are consequences of K. Define a probability function Bel_0 by

$$Bel_0(\alpha) = \frac{c_i}{|S_{\psi_i}|}$$

where i is such that $\alpha \in S_{\psi_i}$. If Bel_0 satisfies K, then Bel_0 is a calibrated solution of K, by the previous theorem, since it is clearly calibrated with respect to atoms.

Interestingly calibration, justified above by appealing to the desirability of accuracy, can also be justified by common sense via the principles of obstinacy and renaming. The following theorem is due to C.Artingstall and J.Paris [2]. Its corollary is stated by P.Lehner in [9].

Theorem 8 *Let $K \in CL$. Then*

(i)   *If K has a calibrated solution Bel_0 then $Bel_0 = N(K)$ for any inference process satisfying obstinacy and renaming.*

(ii)   *If all inference processes satisfying obstinacy and renaming agree on K then this common value is a calibrated solution of K.*

Corollary 1 *If a calibrated solution of $K \in CL$ exists then it is unique.*

NOTE

* Supported by a Science and Engineering Research Council Research Assistantship under the Logic for IT initiative.

REFERENCES

[1] Aczél, J.: 1966, *Functional Equations and their Applications*. Academic Press, New York and London.

[2] Artingstall, C. and Paris,J.: 1991, *M.Sc. Thesis*. Manchester University.

[3] Courtney, P.R.: 1992, *Ph.D. Thesis*. Manchester University.

[4] Cox, R.T.: 1946, "Probability, Frequency and Reasonable Expectation", *American Journal of Physics*, **14**, Number 1, 1-28.

[5] de Finetti, B.: 1937, "La Prévision: ses Lois Logique, ses Sources Subjectives", *Annales de l'Institut Henri Poincaré*, **7**, 1-68.

[6] Horwich, P.: 1982, *Probability and Evidence*. Cambridge: Cambridge University Press.

[7] Kemeny, J. G.: 1955, "Fair Bets and Inductive Probabilities" *Journal of Symbolic Logic* **20**, Number 3, 263-273.

[8] Lehman, R.S.: 1955, "On Confirmation and Rational Betting" *Journal of Symbolic Logic* **20**, Number 3, 251-262.

[9] Lehner, P.: 1991, "Probabilities and Reasoning about Probabilities" *International Journal of Approximate Reasoning* **5**, Number 1, 27-43.

[10] Miller, D.: 1974, "Popper's Qualitative Theory of Versimilitude", *British Journal for the Philosophy of Science* **25**, 166-177.

[11] Maung, I.: 1992, *Ph.D. Thesis*. Manchester University.

[12] Paris, J. and Vencovská, A.: 1990, "A Note on the Inevitability of Maximum Entropy", *International Journal of Approximate Reasoning* **4**, Number 3, 183-224.

[13] Ramsey, F.P.: 1931, *The Foundations of Mathematics and other Logical Essays*. London.

[14] Shimony, A.: 1955, "Coherence and the Axioms of Confirmations", *Journal of Symbolic Logic* **20**, Number 3, 1-28.

Department of Mathematics
Manchester University
Manchester M13 9PL
England

AUTHOR'S INDEX

Acero, J.J., 135
Aczél, J., 231, 259
Allen, S., 114
Almog, J., 156
Aristotle, 165, 195
Artingstall, C. 257, 259
Baker, L. R., 93
Barnden, J., 196
Barsalou, L., 53, 56
Barwise, J., 95, 100-1, 103-5, 110, 113, 114
Berkeley, G.,181-2, 187
Bisiach, E., 32
Black, M., 173, 179
Block, N., 6, 30, 31, 158
Botempo, C. J., 57
Bradley, F. H., 106-7
Brooks, R., 59, 66-7-8, 70
Brown, H., 36, 56
Brugman, C., 164, 179
Burge, T., 6-7, 9, 11-16, 18, 21, 26, 31-32, 122, 126, 132
Byrne, R.M.J., 192, 196
Carnap, R., 54, 56, 115
Carroll, L., 183
Chomsky, N., 31, 40, 56, 165
Church, A., 95, 98, 100, 103, 105, 113, 114
Churchland, P.M., 47-48, 51, 54, 56, 63, 65, 70, 93
Clark, A., 59, 63, 65, 68, 70
Cohen, P. R., 200, 220
Cornman, J. W., 38, 56
Courtney, P. R., 243, 246, 259
Cox, R. T., 231, 259

Crane, T., 4, 27, 31, 33
Crimmins, M., 158
Davidson, D., 95, 105, 113, 114, 153-54, 157n
Davies, M., 1, 4, 6, 14, 23, 28, 30n, 31-2, 60-1, 65, 68n, 70
Davis, S., 56
Desclés, J. P., 179
de Finetti, B., 229, 259
de Kleer, J., 219, 220
Dennett, D., 31, 32
Dretske, F., 158n, 212
Dummett, M., 123, 132n
Erickson, J. R., 183, 186, 196
Euler, L., 181, 183-87, 190-94
Evans, G., 4-6, 10, 32, 60, 70
Fauconnier, G., 162, 179
Fine, K., 113n, 114
Fodor, J., 11, 32, 63, 66, 68n, 70-1, 93n, 156n, 157n, 158, 163-64, 179
Frege, G., 95, 105, 113n, 114, 115, 123, 127-28, 132n, 135, 137-40, 142-45, 147-48, 154, 156n
Fricker, E., 20, 32
French, P. A., 132n, 158n
García Suárez, A., 115
Gärdenfors, P., 159-60, 169, 171, 173, 177n, 179
Gawron, J. M., 196
Geffner, H., 219, 220
Gentner, D., 185, 196
George, A., 31
Glover, J., 57

261

SUBJECT INDEX

268